W9-CPO-286

hKVp

# Succeed with Productivity and Quality

Also available from ASQ Quality Press:

*The Making of a World-Class Organization*
E. David Spong and Debbie J. Collard

*Communication: The Key to Effective Leadership*
Judith Ann Pauley and Joseph F. Pauley

*Proving Continuous Improvement with Profit Ability*
Russ Jones

*Measuring Customer Satisfaction and Loyalty: Survey Design, Use, and Statistical Analysis Methods,* Third Edition
Bob E. Hayes

*Lean for Service Organizations and Offices: A Holistic Approach for Achieving Operational Excellence and Improvements*
Debashis Sarkar

*Managing Service Delivery Processes: Linking Strategy to Operations*
Jean Harvey

*The Executive Guide to Understanding and Implementing Lean Six Sigma: The Financial Impact*
Robert M. Meisel, Steven J. Babb, Steven F. Marsh, and James P. Schlichting

*Lean Kaizen: A Simplified Approach to Process Improvements*
George Alukal and Anthony Manos

*Root Cause Analysis: Simplified Tools and Techniques,* Second Edition
Bjørn Andersen and Tom Fagerhaug

*The Certified Manager of Quality/Organizational Excellence Handbook,* Third Edition
Russell T. Westcott, editor

*Enabling Excellence: The Seven Elements Essential to Achieving Competitive Advantage*
Timothy A. Pine

*Benchmarking: The Search for Industry Best Practices That Lead to Superior Performance*
Robert C. Camp

To request a complimentary catalog of ASQ Quality Press publications, call 800-248-1946, or visit our Web site at http://www.asq.org/quality-press.

# Succeed with Productivity and Quality

## How to Do Better with Less

Imre Bernolak

ASQ Quality Press
Milwaukee, Wisconsin

CARL CAMPBELL BRIGHAM LIBRARY
EDUCATIONAL TESTING SERVICE
PRINCETON, NJ 08541

American Society for Quality, Quality Press, Milwaukee 53203
© 2009 by ASQ
All rights reserved. Published 2009
Printed in the United States of America
15  14  13  12  11  10  09      5  4  3  2  1

**Library of Congress Cataloging-in-Publication Data**

Bernolak, Imre.
  Succeed with productivity and quality : how to do better with less / Imre Bernolak.
    p. cm.
  Includes bibliographical references and index.
  ISBN 978-0-87389-771-6 (hard cover : alk. paper)
  1. Industrial productivity.  2. Industrial management.  3. Technological innovations.
  I. Title.

  HD56.B48 2009
  658.5'15—dc22                              2009021241

ISBN: 978-0-87389-771-6

No part of this book may be reproduced in any form or by any means, electronic,
mechanical, photocopying, recording, or otherwise, without the prior written permission
of the publisher.

Publisher: William A. Tony
Acquisitions Editor: Matt T. Meinholz
Project Editor: Paul O'Mara
Production Administrator: Randall Benson

ASQ Mission: The American Society for Quality advances individual, organizational,
and community excellence worldwide through learning, quality improvement, and
knowledge exchange.

Attention Bookstores, Wholesalers, Schools, and Corporations: ASQ Quality Press
books, videotapes, audiotapes, and software are available at quantity discounts with
bulk purchases for business, educational, or instructional use. For information,
please contact ASQ Quality Press at 800-248-1946, or write to ASQ Quality Press,
P.O. Box 3005, Milwaukee, WI 53201-3005.

To place orders or to request a free copy of the ASQ Quality Press Publications
Catalog, including ASQ membership information, call 800-248-1946. Visit our
Web site at www.asq.org or http://www.asq.org/quality-press.

Printed in the United States of America

 Printed on acid-free paper

 Quality Press
600 N. Plankinton Avenue
Milwaukee, Wisconsin 53203
Call toll free 800-248-1946
Fax 414-272-1734
www.asq.org
http://www.asq.org/quality-press
http://standardsgroup.asq.org
E-mail: authors@asq.org

# Table of Contents

*List of Figures and Tables* . . . . . . . . . . . . . . . . . . . . . . . . . . . . . . . . . *xi*
*Foreword* . . . . . . . . . . . . . . . . . . . . . . . . . . . . . . . . . . . . . . . . . . *xiii*
*Preface* . . . . . . . . . . . . . . . . . . . . . . . . . . . . . . . . . . . . . . . . . . . *xvii*

## Part I    Understanding Productivity

### Chapter 1    What Is Productivity? How to Work Smarter, Not Harder.

**Chapter 1    What Is Productivity? How to Work
Smarter, Not Harder.** . . . . . . . . . . . . . . . . . . . . . . . . . . . . . . . . . **3**
   Introduction . . . . . . . . . . . . . . . . . . . . . . . . . . . . . . . . . . . . 3
   Productivity Is Always Compared with Some Standard . . . . . . 5
   Evidence Proves That Significant Productivity
      Improvement Is Possible and Everyone Benefits. . . . . . . . . . 8
   High Quality Increases Sales, Jobs, and Profits; Poor
      Quality Increases Costs, Lowers Profits . . . . . . . . . . . . . . . 10
   What "Quality" Means. . . . . . . . . . . . . . . . . . . . . . . . . . . . . . 13
   Questions . . . . . . . . . . . . . . . . . . . . . . . . . . . . . . . . . . . . . . 18
   Suggested Answers. . . . . . . . . . . . . . . . . . . . . . . . . . . . . . . . 18

**Chapter 2    Special Productivity Improvement Needs
of Smaller Enterprises and Services, Including
Professionals** . . . . . . . . . . . . . . . . . . . . . . . . . . . . . . . . . . . . . **21**
   Why Is Productivity Vital to Small and Medium-Size
      Enterprises?. . . . . . . . . . . . . . . . . . . . . . . . . . . . . . . . . . . 21
   Measurement of Results Is a Must for SME Managers . . . . . . . 24
   Steps to Improve the Productivity of SMEs. . . . . . . . . . . . . . . 26
   Productivity Needs of Professionals and Others in Services . . . 28
   Measurement Needs of the Knowledge Group. . . . . . . . . . . . . 31
   Examples of Productivity Improvement Ideas for
      Professionals, Other Knowledge Workers, and
      Indirect Workers . . . . . . . . . . . . . . . . . . . . . . . . . . . . . . . 32
   Questions . . . . . . . . . . . . . . . . . . . . . . . . . . . . . . . . . . . . . . 35
   Suggested Answers. . . . . . . . . . . . . . . . . . . . . . . . . . . . . . . . 35

## Part II    How to Identify Productivity Problems and Opportunities: Measurement and Analysis

**Chapter 3    You Need to Know the Facts before
  Taking Corrective Action**. . . . . . . . . . . . . . . . . . . . . . . . . . . . .    **39**
  Objectives and Uses of Productivity and Quality
      Indicators . . . . . . . . . . . . . . . . . . . . . . . . . . . . . . . . . . . . .    39
  Analysis of the Present Methods of Operation. . . . . . . . . . . . .    47
  Questions . . . . . . . . . . . . . . . . . . . . . . . . . . . . . . . . . . . . . . . .    49
  Suggested Answers. . . . . . . . . . . . . . . . . . . . . . . . . . . . . . . . . .    49

**Chapter 4    Developing Productivity Measures** . . . . . . . . . . . . . .    **53**
  Basic Statistical Concepts and the Presentation
      of Data . . . . . . . . . . . . . . . . . . . . . . . . . . . . . . . . . . . . . . . .    53
  Measurement of Quality and Process Control . . . . . . . . . . . . . .    55
  Productivity Measurement Principles and Requirements . . . . .    58
  Steps in Gathering Essential Data . . . . . . . . . . . . . . . . . . . . . . .    62
  Questions . . . . . . . . . . . . . . . . . . . . . . . . . . . . . . . . . . . . . . . .    64
  Suggested Answers. . . . . . . . . . . . . . . . . . . . . . . . . . . . . . . . . .    65

**Chapter 5    Measuring Output and Input** . . . . . . . . . . . . . . . . . . .    **67**
  Criteria and Methods of Choosing Output Indicators . . . . . . . .    67
  Identifying Problems and Opportunities of Output
      Quality . . . . . . . . . . . . . . . . . . . . . . . . . . . . . . . . . . . . . . . .    71
  What Data Are Needed for Measuring Output in Services . . . .    75
  The Measurement of Inputs . . . . . . . . . . . . . . . . . . . . . . . . . . . .    77
  Questions . . . . . . . . . . . . . . . . . . . . . . . . . . . . . . . . . . . . . . . .    79
  Suggested Answers. . . . . . . . . . . . . . . . . . . . . . . . . . . . . . . . . .    79

**Chapter 6    Comparing Productivity Indexes of
  Two Organizations** . . . . . . . . . . . . . . . . . . . . . . . . . . . . . . . . .    **81**
  Methods of Construction . . . . . . . . . . . . . . . . . . . . . . . . . . . . .    81
  Tasks for Practicing Productivity Calculations . . . . . . . . . . . . .    95

**Chapter 7 Benefits of Benchmarking**. . . . . . . . . . . . . . . . . . . . . . .    **107**
  The Importance of Comparative Organizational
      Performance Analysis . . . . . . . . . . . . . . . . . . . . . . . . . . . . . .    107
  The Method of Integrated Organizational Performance
      Analysis . . . . . . . . . . . . . . . . . . . . . . . . . . . . . . . . . . . . . . . .    110
  Performance Improvement Opportunities Revealed
      by Benchmarking. . . . . . . . . . . . . . . . . . . . . . . . . . . . . . . . . .    114
  Questions . . . . . . . . . . . . . . . . . . . . . . . . . . . . . . . . . . . . . . . .    116
  Suggested Answers. . . . . . . . . . . . . . . . . . . . . . . . . . . . . . . . . .    116

# Part III  Basic Elements of Productivity Improvement: Making Personal Efforts More Efficient and Effective

**Chapter 8  Lessons from Successful Productivity and Quality Improvement Programs** . . . . . . . . . . . . . . . . . . . . . . .  **121**

Many Barriers to Productivity Have Been Observed . . . . . . . . .  121

The Productivity Improvement Programs of the Past Had Avoidable Weaknesses . . . . . . . . . . . . . . . . . . . . . . . . . . . . . .  124

Customer Orientation and Feedback . . . . . . . . . . . . . . . . . . . . .  127

Overwhelming Importance of the Human Factor, and Effective Management of Employees . . . . . . . . . . . . . . . . . .  128

Changing Role of Managers and Supervisors . . . . . . . . . . . . . .  130

Questions . . . . . . . . . . . . . . . . . . . . . . . . . . . . . . . . . . . . . . . . . .  133

Suggested Answers . . . . . . . . . . . . . . . . . . . . . . . . . . . . . . . . . . .  133

**Chapter 9  There Are Many Opportunities for Improving Communication** . . . . . . . . . . . . . . . . . . . . . . . . . . . . . . . . . .  **135**

How to Improve Communication within the Organization and with Suppliers, Customers, and Clients . . . . . . . . . . . . .  135

Preparing Efficient and Effective Written Reports and Oral Presentations . . . . . . . . . . . . . . . . . . . . . . . . . . . . . . . .  136

Conducting Productive Meetings . . . . . . . . . . . . . . . . . . . . . . .  137

How to Prepare and Conduct Successful Negotiations . . . . . . .  138

Questions . . . . . . . . . . . . . . . . . . . . . . . . . . . . . . . . . . . . . . . . . .  140

Suggested Answers . . . . . . . . . . . . . . . . . . . . . . . . . . . . . . . . . . .  141

**Chapter 10  Effective Methods of Motivating for Results** . . . . .  **143**

Motivation Is Essential for Enhancing Productivity and Quality . . . . . . . . . . . . . . . . . . . . . . . . . . . . . . . . . . . . . . . . .  143

Nonfinancial Incentives Are Very Valuable Motivators . . . . . . .  144

Involvement and Participation Are Essential Nonfinancial Motivators . . . . . . . . . . . . . . . . . . . . . . . . . . . . . . . . . . . . . .  146

Financial Incentives, Profit Sharing, ESOP . . . . . . . . . . . . . . .  148

Productivity Gainsharing . . . . . . . . . . . . . . . . . . . . . . . . . . . . .  151

Questions . . . . . . . . . . . . . . . . . . . . . . . . . . . . . . . . . . . . . . . . . .  155

Suggested Answers . . . . . . . . . . . . . . . . . . . . . . . . . . . . . . . . . . .  156

**Chapter 11  The Need for Continual Training and Retraining** . . . . . . . . . . . . . . . . . . . . . . . . . . . . . . . . . .  **159**

Training Is Essential for Success in the Twenty-First Century . . . . . . . . . . . . . . . . . . . . . . . . . . . . . . . . . . . . . . . . .  159

Training In Productivity Is a Special Need . . . . . . . . . . . . . . .    163
Effective Training Methods . . . . . . . . . . . . . . . . . . . . . . . . . . .    168
The Role of Productivity Consultants. . . . . . . . . . . . . . . . . . . .    171
Questions  . . . . . . . . . . . . . . . . . . . . . . . . . . . . . . . . . . . . . . .    172
Suggested Answers. . . . . . . . . . . . . . . . . . . . . . . . . . . . . . . . .    173

## Part IV    How to Improve Productivity through Organization and Technology

### Chapter 12    You Can't Succeed without a Plan . . . . . . . . . . . . .    177
The Need for Setting Objectives and Formulating
    Productive Strategic and Operating Plans . . . . . . . . . . . . . .    177
Key Characteristics of Successful Plans. . . . . . . . . . . . . . . . . .    180
What Should Be Included in a Business Plan and How to
    Develop a Plan  . . . . . . . . . . . . . . . . . . . . . . . . . . . . . . . . . .    181
Scheduling . . . . . . . . . . . . . . . . . . . . . . . . . . . . . . . . . . . . . . . .    185
Decision Making and Problem Solving. . . . . . . . . . . . . . . . . . .    189
Questions  . . . . . . . . . . . . . . . . . . . . . . . . . . . . . . . . . . . . . . .    189
Suggested Answers. . . . . . . . . . . . . . . . . . . . . . . . . . . . . . . . .    190

### Chapter 13    How to Design Productive Operations of High Quality . . . . . . . . . . . . . . . . . . . . . . . . . . . . . . . . . . . . .    193
Setting Priorities for Areas of Potential Productivity
    Improvement. . . . . . . . . . . . . . . . . . . . . . . . . . . . . . . . . . . . .    193
Customers and Clients Now Want Proof of Quality  . . . . . . . . .    194
How to Ensure High-Quality Production and Service . . . . . . . .    197
Great Opportunities in Administration  . . . . . . . . . . . . . . . . . .    200
Product Design, Standardization, Simplification, and
    Specialization. . . . . . . . . . . . . . . . . . . . . . . . . . . . . . . . . . . .    204
Purchasing, Plant Organization, and Inventory
    Management. . . . . . . . . . . . . . . . . . . . . . . . . . . . . . . . . . . . . .    206
Productivity Improvement in Physical Distribution . . . . . . . . . .    209
Questions  . . . . . . . . . . . . . . . . . . . . . . . . . . . . . . . . . . . . . . .    211
Suggested Answers. . . . . . . . . . . . . . . . . . . . . . . . . . . . . . . . .    211

### Chapter 14    How to Make the Best Use of Your Time, Efforts, Knowledge, and Other Resources . . . . . . . . . . . . . . .    215
Utilize Your Time Effectively. . . . . . . . . . . . . . . . . . . . . . . . . .    215
Apply Productive Methods of Organization, Work Design,
    and Job Assignment. . . . . . . . . . . . . . . . . . . . . . . . . . . . . . .    216
Control Is Necessary for Achieving Productivity. . . . . . . . . . . .    219
Personnel Controls and Accountability. . . . . . . . . . . . . . . . . . .    222

How to Benefit from Technological Progress . . . . . . . . . . . . . . .     225
Questions . . . . . . . . . . . . . . . . . . . . . . . . . . . . . . . . . . . . . . . .     228
Suggested Answers . . . . . . . . . . . . . . . . . . . . . . . . . . . . . . . .     229

*Endnotes* . . . . . . . . . . . . . . . . . . . . . . . . . . . . . . . . . . . . . . . .     *231*
*Index.* . . . . . . . . . . . . . . . . . . . . . . . . . . . . . . . . . . . . . . . . . .     *235*

# List of Figures and Tables

Figure 4.1    Illustration of a Pareto analysis. . . . . . . . . . . . . . . . . . . . . . . .    56

Figure 4.2    Illustration of an $\bar{X}$ chart. . . . . . . . . . . . . . . . . . . . . . . . . . . .    57

Figure 6.1    Productivity differences in percent from absolute data, homogeneous output. . . . . . . . . . . . . . . . . . . . . . . . . . . . . .    82

Figure 6.2    Construction of productivity index from raw productivity data, homogeneous outputs. . . . . . . . . . . . . . . . . . . . . . . . .    84

Figure 6.3    Productivity index constructed from indexes, homogeneous outputs. . . . . . . . . . . . . . . . . . . . . . . . . . . . . . . . . . . . . . . . .    86

Figure 6.4    Basic data for calculating weighted productivity indexes over time. . . . . . . . . . . . . . . . . . . . . . . . . . . . . . . . . . . . . . .    87

Figure 6.5    Calculating ULR weights (ULR = unit labor requirement in year 1). . . . . . . . . . . . . . . . . . . . . . . . . . . . . . . . . . . . . . . .    88

Figure 6.6    Calculating weighted output indexes over time. . . . . . . . . . .    90

Figure 6.7    Calculation of input indexes over time. . . . . . . . . . . . . . . . . .    91

Figure 6.8    Calculating productivity indexes from weighted output and input indexes. . . . . . . . . . . . . . . . . . . . . . . . . . . . . . . . . . .    92

Figure 6.9    Productivity of two producers—heterogeneous outputs, ULR weights. . . . . . . . . . . . . . . . . . . . . . . . . . . . . . . . . . . . . .    95

Figure 10.1   Scanlon bonus computation for company X. . . . . . . . . . . . . .    154

Figure 10.2   Company X bonus report for January (current year). . . . . . .    155

Figure 12.1   Basic Gantt chart. . . . . . . . . . . . . . . . . . . . . . . . . . . . . . . . . . . .    187

# Foreword

A seminal piece of work. I wouldn't have expected less from Imre. He has delivered one of the best, most readable and comprehensive pieces of work on productivity I have ever read.

To phrase it in his mind-set, *you* are the customer for this book, and he writes to you—and regardless of your sophistication and knowledge about productivity, there is something in here for you. There was for me.

As a professor at four universities, and dean of two business schools, I've always known in a general sense about the importance of productivity and quality. But it wasn't until I was chairman of the U.S. Price Commission from 1971–73, the period of price and wage controls under President Nixon, that it was brought home to me how really important productivity was to offset wage-driven inflation. In fact, I required all those who wanted a price increase to write on their request for a price increase what their productivity growth was as an offset to increasing prices.

What astounded me was how ignorant the biggest and best American firms were about productivity—how to measure it and the role it played in their organizations and their ability to compete internationally. They were also way behind their international competitors in productivity and quality—which they assumed at that time was just not true.

That was why, after I left the government, I started the nonprofit American Productivity Center (APC) in 1975. In fact, Imre assisted us in that effort, and particularly with his work on interfirm comparisons of productivity—something that was almost unknown among American firms, but which Imre had done for years with Canadian and other international organizations.

Later, I changed the name of the Center to the American Productivity and Quality Center (APQC), for as Imre says in Chapter 1, "high quality increases sales, jobs, and profits; poor quality increases costs." Productivity and quality go together. Alan Watts described it graphically: "You can never have the use of the inside of a cup without the outside. The inside and the outside go together."

Many books on productivity are dull, dense, abstruse, technical (often boring), and I've got lots of such books on my shelves. They are barely underlined or marked up, as I'm accustomed to do.

Not this book. The manuscript he sent me is now messy, with my frequent exclamation points, asterisks, and sidebar comments like "amen"—"well said"—"way to go." When the first astronauts stood on the moon, the important thing they said was not that they had set foot on the moon, but that it was the first time they had set eyes on earth. This book gave me the same feeling about productivity.

If you can't tell by now that I really liked this book, you aren't listening. I won't recite the contents. The table of contents does a better job than I could. But let me tell you the things I especially liked.

One, he meets head-on in every chapter the misconceptions about productivity. He's right that many people mistakenly assume they know all about productivity, or that it's something to be feared and hated. If they would just read this book, almost every chapter would mitigate these fears, and might even make them converts. I'm a dreamer here, but we have to keep trying!

Two, right up front, in Chapter 2, he gives examples and practical help to smaller enterprises. I like that because too many assume that productivity is only for the larger firms and only for the manufacturing sector. He tackles both issues in Chapter 2 and also in later chapters showing that productivity techniques and simple measures can also be used by small firms and by service organizations. And that's very important in today's world of a dominant services sector and a huge and growing number of knowledge workers.

Three, I like the measurement chapters, because this is what trips up lots of people—me included—in the real world of assigning measures and metrics to operations when many organizations don't keep good records, especially time records. He tackles each of the measurement issues, giving concrete examples at the ends of the chapters.

Fourth, I was tremendously impressed with the time and space he gives in Chapters 8 through 11 to the human element in productivity, so often ignored, overlooked, or handled badly. We have inherited in too many organizations "Industrial Age Management" where the employees are considered interchangeable parts and variable costs—not involved, not consulted, not incentivized. And they are organized into Industrial Age dysfunctional silos that breed waste and poor quality. No wonder productivity lags in those firms. And it's why Chapter 11 stresses the need for continual training and retraining. It's ever more essential for the twenty-first century,

All of this fittingly leads to the last few chapters that stress how productivity can be enhanced through planning, organization, and use of time, knowledge, technology, and resources.

Probably one of the most profound statements Imre made in the book that doesn't sound profound at all (a trick of wise people) is that productivity is really simple. I think Archimedes or Copernicus must have said the same thing when they understood what they suddenly knew. Why don't others get it! Only he didn't run naked in the streets or urge people to stand out of his light, he wrote a book.

And a good book it is.

Dr. C. Jackson Grayson
Chairman and CEO,
American Productivity and Quality Center

# Preface

It was recognized by Adam Smith already in the late eighteenth century[1] that people's lives can only be made better and less strenuous by increasing their productivity, that is, by producing more and better with less effort and resources. This knowledge did not become widespread until much later. Although the United States Department of Labor started to measure productivity before the turn of the nineteenth century,[2] the importance of productivity had only become more widely known in the middle of the twentieth century. This is when people started to analyze the economic factors of productivity that can be used to make life better and easier.

The author of this book has been among the earliest and most effective contributors to productivity analysis and improvement in North America, Europe, and Asia. I have built up a half century of experience in the field of productivity analysis and improvement.

The author was among the first in North America to analyze and understand how productivity can be improved. I cooperated with the Productivity and Technology Division of the United States Bureau of Labor Statistics, the American Productivity and Quality Center, particularly in its formative stages, the U.S. Network of Productivity and Quality Centers, as well as productivity organizations in Central and South America. In Europe, I served for over a decade as the director for Canada of the European Association of National Productivity Centres.

In cooperation with these organizations, I directed the development of productivity work and interfirm productivity comparisons in Canada and internationally with the United States and Europe. By the time of this writing, the interfirm comparisons covered more than 4000 companies in over 100 different industries. I also directed such interfirm productivity comparisons throughout Asia as chief expert and editor for the Asian Productivity Organization and, in the 1990s, I served on the staff of the World Bank to improve productivity in Barbados. I authored many productivity-related publications in America, Europe, and Asia, including many reports, book chapters, and articles on productivity analysis and improvement methods, published by the Organisation for Economic Co-Operation and Development (OECD),[3] the National Bureau of Economic Research,[4] the American Productivity Center,[5] the British Council of Productivity Associations,[6,7,8] the American Institute of Industrial Engineers,[9] the Asian Productivity Organization,[10,11] the Canadian Bureau of Management Consulting,[12] the Washington-based International Productivity Service,[13] and the European Association of National Productivity Centres.[14]

The productivity analyses and interfirm comparisons showed that virtually every organization, even the best, can learn from their competitors and counterparts, as well as from self-analysis, about how to achieve more and better through improved organization and utilization of their resources. This book describes and shares the lessons learned from these decades of productivity analysis and improvement work.

This basic and comprehensive book is intended for entrepreneurs, managers of local branches of large corporations, such as banks or business chains, as well as managers or aspiring managers in other private or public organizations. It is essential reading for students of business administration and economics, as well as managerial practices, and fills a hole in the training of students in all fields where they will manage people and resources. Professionals, other knowledge workers, and technical people also benefit because their professional training usually concentrates on their specific expertise and not productivity improvement. Over the years it has become clear that even managers of the best organizations can benefit by learning from the experience of others.

The book contains four parts. Part I explains what productivity is and why it's so important. Part II describes how productivity problems and opportunities can be identified through measurement and systematic analysis. While this is not a statistical textbook, it explains through simple and practical solutions how one can benefit from relevant measurement. Part III outlines how each individual person can improve their productivity and become significantly more efficient and effective. Part IV reviews how productivity can be enhanced through better planning, organization, and use of time, knowledge, technology, and resources.

In summary, this book will show you how you can do more and better with less effort and resources.

Imre Bernolak
Dr. Political Science (Economics), LLD

# Part I

## Understanding Productivity

# 1

# What Is Productivity?

*How to Work Smarter, Not Harder*

## INTRODUCTION

This is a practical book. It brings together thousands of ideas and experiences of outstanding managers of private and public organizations in North America, Europe, and Asia. It describes how these managers used their methods and approaches to successfully manage their own work and that of their organizations. This book explains what productivity is and what quality is, how they help each other, and why each is very important to everyone. It examines how to identify, evaluate, and improve the relevant problems and opportunities. Higher productivity ensures competitiveness, opportunities, and jobs.

What is productivity? Who benefits from its improvement? Who can improve it and how? The word *productivity* is used very often but very few people understand what it really means, how vitally important it is to everyone, and that it can be improved dramatically if we learn how to do it.

Productivity is often taken to mean "'production" or "performance," although "production" or "performance" means simply how much we produce or provide while "productivity" expresses how much we produce or provide *per* resources used. Another source of misunderstanding is that values increased by inflation are presented as productivity growth even though the real volume of goods or services may not have grown at all, or perhaps even declined. We may pay more for our groceries this year than last year but this does not necessarily mean that we are getting more or better food than last year.

It is often assumed that increased productivity is brought about mainly by higher technology, larger scale of operations, or harder work. Later on we shall examine the most important requirements of productivity improvement, but it needs to be pointed out here that while the technology and equipment used are important factors of productivity, and while the scale of operations may be significant, careful and systematic planning, proper organization, training, motivation, communication, staff involvement, incentives, and "smart" work are usually far more important determinants of productivity levels and growth than technology or scale.

It is also absolutely wrong to assume that high productivity means poor-quality, sloppy work. Just the opposite! We shall see later that high quality is an essential factor of high productivity. Evidence also shows that it is incorrect to assume, as it is frequently done, that the productivity of the rapidly growing sector of professional and other knowledge workers, or indirect workers, can not be measured. It has been done and is being done. As a result of such misconceptions it is not sufficiently recognized in the operational reports of either the private or public sector. Business enterprises and public organizations seldom report what their actual productivity results have been, and meaningful productivity efforts often fail to earn sufficient reward.

The basic meaning of productivity is simple. It means what and how much we produce with our efforts from the resources we use. We work in order to produce the goods and services that we want. Productivity is the *relationship* between *what is produced* and the *resources used* in its production. Productivity is output per input, such as $x$ number of chairs produced per $y$ number of labor hours.

$$\text{Productivity} = \frac{\text{Output}}{\text{Input}}$$

There are, however, many ways of looking at productivity. The productivity of a worker may differ due to many factors such as his or her ability and effort, the tools available, the organization of the work, and so on. Productivity is like health. It has many determinants and must be viewed from many angles to understand it and be able to improve it. When we do not feel well, the doctor will look at the symptoms, take our temperature, order tests and X-rays, ultrasounds, or MRIs, and so on, and from all these findings he or she can choose the medication needed to make us better. It is the same with productivity, our economic health. It is necessary to look at what resources are used and how they are combined to produce the results we want.

Productivity, therefore, consists of a family of concepts and measures. One measure of productivity shows how much a person can produce in a certain time period with the available machines and tools. For instance, a worker can produce a certain number of chairs per day. Another productivity measure shows how much can be produced from the materials used, for example, how many chairs can be made from a certain amount of wood. A third productivity measure shows how many products can be manufactured with a kilowatt-hour of energy. The better we combine and use the resources, the higher our productivity will be and the better off we will be.

The result of productivity is the "pie." The only way we are going to get a bigger "pie" for the work we do for ourselves, our families, or the organizations that provide our jobs is by increasing the pie. Productivity is not the same as profitability although in the long run productivity determines profitability. In the short term, however, profits are also influenced by many other factors such as current market conditions, inflation, changing tastes, and fads. An item in high demand and short supply may lead to high prices and sudden changes in profits. Since productivity is, however, a volume relationship between outputs and inputs, it is not influenced by supply/demand fluctuations, consumer preferences, or inflationary changes.

## PRODUCTIVITY IS ALWAYS COMPARED WITH SOME STANDARD

Productivity by itself means very little. It hardly means anything to say that so many units were produced per unit of resource input without being able to compare it with some standard. Productivity is a relative concept. We may compare a productivity relationship with a similar relationship somewhere else or with a standard at the same point in time. This is called a *cross-sectional* productivity comparison. For example, one company is 25 percent more productive than another, or the performance of one company unit is 20 percent less than the given standard for that time.

Alternatively, we can compare the input–output relationship with some historical standard: "Has the productivity ratio improved over time?" This is called a *longitudinal,* or *time,* comparison and is usually expressed in index form. It can show, for instance, that productivity of a person, an office, an operating unit, factory, or machine increased by five percent from last year.

If we measure over time the change of values rather than physical units, the values must be expressed in constant-value terms, which exclude the impact of inflation. If, say, the output expressed in dollars went up by 10 percent but the prices of the same product or service increased in the meantime by five percent, the constant-dollar or "real" output increased only by five percent rather than by 10 percent. Physical unit counts or other volume measures are preferable but very often a variety of products are produced by the same person(s) or the same machine, and can only be added up in value terms. Then these must be expressed in constant dollars. (Different outputs can also be expressed and added up in "unit-labor-requirement" [ULR] terms but such "weights" are not available too often. The procedure of combining different units by "weights" will be discussed later on, in the productivity measurement section).

There are two basic types of productivity concepts and measures, namely the broader "effectiveness type" concepts and measures, and the narrower "efficiency type."

*Effectiveness*-type measures aim to assess whether an activity or a program makes the best use of efforts and resources (inputs) to achieve its goals (outcome). They show how well an activity or program accomplishes its stated purpose. The efforts and resources used are the "inputs," and the results are the "outcome." Consumer representatives and clients, such as hospital patients, usually prefer the effectiveness-type measures, such as surgery good outcomes per doctor. What good does it do for a patient if the activities were efficient, that is, used the least resources to produce the output, but were ineffective in producing the desired results.

*Efficiency*-type measures, on the other hand, compare the final goods or services a person or an organization produces with the resources used. They relate a given resource use or activity to its direct results, which consist of some kind of finished product produced or service provided. Efficiency measures assess how well the job is done in relation to the consumption of labor time, capital, materials, and/or other resources used. For example, more beds are being made up properly by a new method than before. These measures do not attempt to determine whether the outputs should be produced or whether the inputs do achieve some desired goal.

The main difference between the two types of measures is in the definition of output, namely that efficiency-type measures relate inputs to direct *outputs* while effectiveness-type measures relate inputs to *outcome*, that is, end results, impacts. Another way of looking at the two aspects of productivity is that efficiency looks at using resources "the right way" while effectiveness focuses on doing "the right thing."

For example, how many drunk drivers were caught by policeman per day is an efficiency-type measure while how many less accidents are caused by drunk drivers per policeman on duty is an effectiveness-type measure. The difference between the two types of measures is more significant in the public sector than in the private sector, because in the private sector the product or service provided is the final output demanded and purchased in the market. In the public sector the final result of a service may be far beyond the final output of an agency and may be much more difficult to measure because of the involvement of utility and value judgements.

In either type of productivity measure, it is essential to clearly define the output to measure. In effectiveness-type measures it is also important to critically examine the causal relationship between the program output and program result or impact. In the case of hiring, for instance, the valid efficiency measure would be the number of persons hired in a year per personnel officer, while an effectiveness measure could be the number of those employees still on staff, say, a couple of years later, over the total number hired per personnel officer. Most of the effort should focus on the development of efficiency-type productivity measures, which yield relatively better payoff by helping to show whether the organization's output is produced efficiently.

In view of the basic importance of people with regard to productivity both as producers and as consumers, the *labor productivity* concept has been the preferred expression of productivity for many years. It is a measure of how much of some goods or services of a certain quality are produced per person-year or per person-hour. This relationship, usually expressed as the ratio of output (that means the goods and services produced) per labor input (worker hours or years) shows how well an organization utilizes and converts all of its resources—manpower, materials, equipment, capital, and energy—into outputs, that is, tangible items or services, measured per unit of labor time.

$$\text{Labor productivity} = \frac{\text{Output (goods or services produced)}}{\text{Labor input (worker hours or years)}}$$

Although these measures relate output to labor input, they do not measure the specific contribution of labor or any other factor of production. Rather, they reflect the joint effects of a number of interrelated influences, such as technology, capital investment per worker, capacity utilization, level of output, plant layout, managerial and supervisory skills, as well as the skill, interest, and effort of the workforce. The factors of productivity and the means of productivity improvement will be discussed later, but it needs to be emphasized here that productivity improvement is not about making people work harder but rather "smarter" by organizing, motivating, and equipping them with better tools to help them work more efficiently and effectively.

As people have realized that productivity is not only affected by labor but also by other factors, they endeavored to find out the relative importance of these factors in order to develop the best combination of all resource inputs toward achieving higher productivity. They have developed "output per capital" productivity measures, "output per material" productivity measures, "output per energy" productivity measures, and so on. Each of these may be needed for their specific purpose. Since these measures relate output to one input only, they are called *partial productivity* concepts and measures.

It has been considered, however, that productivity improvement often requires heavy capital expenditures and, therefore, it would be useful to have a measure that shows the net productivity increase of the combined labor and capital inputs. *Multifactor* productivity measures have, therefore, been developed that show the *ratio of output* to the *combined input* of several inputs, such as labor and capital.

$$\text{Multifactor productivity} = \frac{\text{Output}}{\text{Combined inputs}} = \frac{\text{Output}}{\text{Labor input} + \text{Capital input}}$$

The multifactor measures show whether the higher "labor productivity" has been brought about by adding more machines or facilities to assist labor at extra cost—which perhaps offsets the labor saving. Multifactor productivity measures are particularly important in capital-intensive enterprises. Conversely, *total factor productivity* (TFP) measures, which relate output to the combination of all inputs, are difficult to measure and interpret and are rarely used in individual organizations.

Productivity ratios also vary depending on what output is examined. In the case of measurable physical outputs, such as shoes produced or cheques issued, the number of shoes made or cheques issued can be measured per unit of labor input or the combination of labor and capital inputs. If more shoes of the same type are produced from the same resources or from the same combination of resources, higher productivity has been achieved. The number of cheques issued, error-free lines typed, or drawings made per employee-hour are other examples of productivity measures based on physical count. Productivity measures can also be usefully combined with profitability data. These combinations will later be discussed in detail.

# EVIDENCE PROVES THAT PRODUCTIVITY IMPROVEMENT IS POSSIBLE AND EVERYONE BENEFITS

Information received from large numbers of private firms and associations as well as public agencies in the United States, Canada, Europe, and Asia provides overwhelming evidence that productivity can be improved dramatically provided that planned productivity actions are taken or productivity programs are implemented. There is no such thing as "instant success," but the growth of organizationwide management improvement efforts in large organizations has been rapid, and substantial results have been reported, starting with the first year. Similarly, there is overwhelming evidence that smaller organizations can also achieve high productivity if proper approaches are used.

For example, the Canadian Interfirm Comparison Program that the author introduced, developed, and directed for more than 20 years, which covered more than 4000 companies in about 100 manufacturing and service sectors in North America and elsewhere, has shown that 40 percent variation in productivity is quite common between the average and best comparable firms. In several industries, the spread between the average and best firms ranges up to 100 percent. If productivity is compared among the worst and best comparable competitors, it varies up to 300 percent and more. It should be emphasized that these firms consider themselves comparable with each other and competing with each other. Virtually all have been happy with the comparisons because, on a strictly anonymous basis, they showed each of them the areas where they could improve their productivity. A large number of companies with successful productivity improvement programs also reported that their *annual* productivity improvements reached five percent or even higher.

Increasing productivity lies at the heart of building and sustaining economic success. Productivity, however, is not one of those things that can be forced on people. In order to really try to improve productivity and seize every opportunity to contribute to its enhancement, everyone needs to understand the fact that productivity improvement is in his or her own best interest. When increased office productivity results in an improved quality of output, such as a reduced number of errors, the customers or clients are happier, the workers are more pleased with their achievement, the employer saves extra costs of having the work redone, and so on.

The significance of productivity improvement to employers is most clearly evident through reduced costs. They benefit from a more productive, stable as well as more satisfied, motivated, and better trained labor force, less absenteeism and turnover. They build new goodwill and reputation. Communications improve. Through productivity measurement the organizations can prepare more accurate and realistic budgets and long-range plans and avoid, for instance, costly incorrect investments. Physical assets, office space, and other kinds of infrastructure are better selected and utilized. Unnecessary activities are eliminated or at least reduced. For example, photocopying tends to be overused and can be eliminated in many situations. Security can be enhanced. Productivity and quality improvement efforts have also been noted to enhance business image and credibility.

Customers benefit when their suppliers or service providers better meet their requirements with higher-quality products and services. They gain if the products and services they want are more available, with reduced response times, and at lower prices. For workers, productivity improvements tend to improve the quality of working life, make the job easier, simpler, more satisfying, more pleasant, and more comfortable with less boredom, reduced task difficulties and distractions, as well as safer with fewer work injuries. The changes result in a better work environment, a more attractive workplace with more suitable furniture, better lighting, reduced noise, and fresher air. Better layout of the workplace can reduce walking distances on the job.

Productivity tends to lead to better pay, improved training opportunities, upward mobility, and better career development. For most workers, productivity improvement means more-secure jobs because improved productivity strengthens and preserves the company and this, in turn, preserves jobs. In other words, productivity today does not mean rushed and overworked assembly line workers, but the best use of all resources for the benefit of all concerned.

## HIGH QUALITY INCREASES SALES, JOBS, AND PROFITS; POOR QUALITY INCREASES COSTS, LOWERS PROFITS

Today's consumers, whether they are personal or institutional consumers, are better educated and informed. They know how to research, how to choose, what characteristics they want in a product or service. They are no longer willing to put up with inferior products or inadequate services. They are no longer satisfied to have merely "safe" products but also demand environmentally acceptable and personally beneficial products and services. Customers today want to know how long the merchandise has been in stock or on the market, and pay less for goods that are not the most recent.

In other words, today's consumers want *value* for their money. As society has developed, the demanded quality attributes have also changed. Price and location are no longer necessarily the most important determinants of the customer's choice. At the same time, the choice of competitive products and services offered to the consumer has become increasingly wider from both at home and from around the globe.

The orientation toward serving customers appears natural when companies start small. Later, however, as the companies grow, assembly lines tend to be set up, work simplification introduced, and functions fragmented. The emphasis tends to become placed on "how" things are done, rather than "why." Customer service becomes a specialized function rather than the concern of all employees. This development builds distance between the producers and the customers and creates conflicts between functional departments. Quality suffers because the customer's interest is no longer the concern of every worker.

In view of the critical importance of the customers or users, all economic endeavors must be reoriented to serve their needs and wants. In order to meet their growing demand for quality, more and more successful organizations are indeed finding ways to organize their work accordingly, enabling them to radically improve the quality of their products and services by focusing on what the customer wants. Producers listen to their customers and endeavor to meet the customers' "valid" requirements and produce products or provide services at the degree of quality demanded by the customers at a given price.

The critical importance of producing goods and services of high quality and focusing on the qualities that are valued by the consumers has been emphasized by research regarding highly successful companies. We have found that companies that sell high-quality products or services are generally more profitable than those whose products or services are of lower quality. In service industries, it has also been recognized that service quality is the key to organizational excellence and success. In banking, for instance, quality of the services provided is the key that differentiates successful firms in the market. These banks provide excellence in service delivery and maintain outstanding customer relations in addition to basic product features, add-on values, contained costs, and competitive prices.

The developments over the years in the rapidly expanding global markets, intensifying competition at home and abroad, and increasing consumer expectations everywhere have made quality a top strategic issue in both domestic and international trade. Obviously, product prices remain very important, but quality considerations have become primary determinants of value. For reasons that will be discussed below, quality factors are now often actually placed ahead of prices in making purchasing decisions.

Examples from many leading companies show that today quality plays a pivotal role in determining competitiveness and market share. For example, Microsoft, Xerox, Hewlett-Packard, General Electric, IBM, Toyota, and Procter & Gamble have substantially improved their competitive advantage through quality improvements. Quality changes enabled them to become more productive and provide better products and services at relatively lower cost. They have proven that quality and productivity can rise together, that it is not necessary to sacrifice one for the other. Rather, they are complementary and mutually supporting aspects of production. Quality improvement aims at finding ways to improve the quality of the product or service provided, while productivity focuses on finding methods to improve the utilization of the required resources.

It costs money to produce high-quality products and provide high-quality services. However, low-quality output usually costs much more. The *cost of quality* is made up of the price of nonconformance to quality standards (the cost of repairs, and so on) and the price of conformance (that is, the cost of preventing the errors and defects). Research has found that the price of nonconformance (cost of repairing, fixing) was in observed cases more than double the price of conformance (doing a quality job the first time). It has also been proven that the overall cost of quality can be reduced significantly.

Quality problems cause immediate nonconformance costs to the organization. Defects and errors are not free. First, those who make them are not only paid but also waste materials in the production of the defective product. When the job has to be redone, the material may have to be bought again, the worker has to be paid again, and the product has to be retested. Faulty products increase the need for inventory, increase the work in process, and cause extra material handling. Defects and errors tie up tools, production and other capital equipment, and reduce the capacity utilization rate. Breakdowns cause costly downtime. The replaced products have to be redistributed. Finally, defects and errors cause significant warranty and liability costs.

In offices, a study of a specific organization found that nearly a quarter of their "intake products," for instance, interview notes and formal charges, were defective. The documents had to be reviewed and then the faulty documents had to be reworked. In order to correct this situation, quality standards were introduced with the result that the proportion of faulty products fell dramatically. In general, studies have indicated that as much as 25 percent of manufacturing and administrative time or more is typically diverted to repairing defects, correcting errors, and apologizing to customers.

As a result of such findings, modern management has increasingly rejected the conventional industry standards traditionally used in the past of "acceptable" quality levels of one or two percent defective. These were usually the minimum degrees of quality needed to remain competitive. Now, however, *zero defect* strategies have become more and more standard in order to surpass or at least match competitors. The quality strategies must be meaningful and effective, practical in implementation, and not mere slogans.

Studies have also indicated that the above-mentioned damages caused by defects and errors are not the end of the problem. If a customer is unhappy with the product or service, only one in 25 tends to complain to the producer—unless there is a complete breakdown during the warranty period. The others will probably stop dealing with that company. In addition, each of the unhappy but "quiet" customers will tell 10 of his or her friends and relatives of the negative experience and also scare them away from the company's products and services. This means that every time a company receives a complaint, nearly 250 other people will be warned against the company's products or services.

The majority of complainers will continue to deal with the company if the problem is resolved promptly. Nevertheless, the customers whose complaints are resolved will only tell very few others about their experience. Because of this it is necessary to examine closely what quality is and why it has become a question of survival for business, as well as a key to jobs for workers. Later on we shall discuss how to identify and measure quality problems, and how quality can be improved.

## WHAT "QUALITY" MEANS

The essential starting point for a quality improvement program is to develop a working definition of quality. What does quality mean? How good is "good"? Is there a general standard of high "quality"?

In basic terms, quality is the degree to which a product or service meets the expectations, requirements, and/or specifications of its users. The producers of goods and the providers of services must focus on what the customers, clients, or patients value, and formulate their strategies accordingly. "Customers" in this sense also includes the internal users of goods or services, such as corporate management, other departments, or their workers, as users of the organization's personnel, legal, accounting, or health services.

The foundation of producing high-quality products or services must be a clear plan in which the intended products or services of high quality are defined. Then the customers or target markets are identified, as well as those factors about the products or services that the customers value. While the above considerations provide a general understanding of the concept of quality, there is no simple or generally applicable definition of quality. It is a complex matter. It consists of a set of characteristics for which a product or service is designed, produced, or provided.

In a technical sense, *quality* is the degree to which a product or service conforms to prescribed specifications. However, quality characteristics that are defined in the specifications vary from industry to industry, product to product, and service to service. The characteristics and the relevant standards need to be specified for each product, subproduct, and service, based on the intended purpose of the product or service, the target markets, and customer requirements.

In view of the different nature of the various quality characteristics, it is not only helpful but essential to make a distinction between *attribute* characteristics and *variables* characteristics. Attribute characteristics are either "good" or "bad." In other words, they either conform to the prescribed standard or do not. For example, a service may be friendly or not friendly. Customer service is either quick or not. There is no degree of conformance regarding attribute characteristics, and the defects can be counted. On the other hand, variables characteristics, such as weight, dimension, and chemical composition, are measurable and vary in their degree of conformance.

Some of the characteristics can be looked at as either an attribute characteristic or a variables characteristic. For example, "power" of a car can be regarded from the customer's viewpoint as "powerful" or "not powerful," an attribute characteristic. On the other hand, from a technical point of view, "power" can be expressed in terms of horsepower, which is a variables characteristic. In another example, "acceleration" may be termed as an attribute characteristic as "fast" or "sluggish," or as a variables characteristic from the technical point of view as "seven seconds from standing to 100 km/hour."

Quality characteristics may also be classified as *objective* or *subjective*. Both affect the value of the product or service. Standards can be set for each, and quality can be defined in terms of conformance of the product or service to those standards. Some of the quality characteristics relate to the product itself while others relate to the services connected with the product. Examples 1-A and 1-B show what characteristics are often considered as important quality indicators.

The desired attribute characteristics and variables characteristics, as well as the objective and subjective quality standards, should be defined, tracked, and upgraded where possible for your products and services. It is important to remember that the quality requirement does not only apply to the end product, because the quality of the end product depends on the quality of the raw materials, components, and all other intermediate inputs along the production line, including managerial, administrative, and other office work.

In other words, as shown in Example 1-C, it is important to consider the quality of any input, direct (labor, raw materials, and so on) or indirect (management, administration, and so on) into the production and delivery of a product or service. Otherwise, the quality of the final goods or services will be inferior. As the critical importance of the high quality of products and services has become recognized over recent years by both individuals

## EXAMPLE 1-A

### Examples of Quality Attributes of Cars

*Objective* quality attributes of cars include—among others—the following:

- Power
- Reliability
- Safety
- Stability
- Maneuverability
- Speed
- Mileage
- Acceleration
- Features
- Finish
- Roominess
- Frequency of service needed
- Frequency of repair needed
- Timely delivery
- Availability of repair
- Prompt response for service

*Subjective* quality attributes of cars include—among others—the following:

- Styling
- Color
- Confidence
- Enjoyability

### Examples of Quality Attributes of Airlines

*Objective*

- Safety
- Availability
- Reliability
- Punctuality
- Ease of check-in
- Seating dimensions
- Cabin cleanliness
- Food (freshness, variety)
- Drinks (choice, availability)

*Subjective*

- Attractive ambience
- Flight attendants' attitudes
- Courtesy
- Carefulness
- Sense of hospitality
- Friendliness
- Attention of staff
- Caring attitude
- Visibility of staff

and organizations, widespread interest has developed in how it can be achieved. Answers to this question will be addressed later, in Chapter 13, when we'll discuss how to design productive operations of high quality.

## EXAMPLE 1-B

### Examples of Quality Attributes of Banks

*Objective*

- Length of lines
- Frequency of computers down
- Interest rates
- Service charges
- Accuracy of statements
- Promptness of statements
- ATM breakdowns
- Add-on value services

*Subjective*

- Helpfulness of staff
- Familiarity of staff
- Continuity of staff
- Prompt return of calls
- Not being left on hold
- Convenience of parking

### Examples of Quality Attributes of Hotels/Motels/Restaurants

*Objective*

- Location
- Rating
- Environment
- Accessibility
- Safety
- Reliability
- Availability of services
- Propriety of housekeeping
- Order and cleanliness
- Business services
- Food (availability, variety, taste, and so on)
- Drinks (availability, choice, and so on)

*Subjective*

- Reputation
- Confidence
- Comfort
- Hospitality
- Attention to requests
- Friendliness to guests

## EXAMPLE 1-C

### Examples of Internal Managerial and Administrative Quality Attributes

*Primarily objective* characteristics

- Clear objectives
- General clarity of instructions
- Effective customer (user, client) interface
- No overlapping or contradictory functions
- Accuracy (no errors, defects)
- Dependability
- Effectiveness (objective, no redundant or unnecessary elements)
- Simplicity (no overcomplexity)
- Timeliness (no delays, delivery on time)
- Turnaround time
- Time needed for correction or repair
- Receiving and processing orders correctly
- Correct deliveries and invoicing
- Prompt follow-up of complaints
- Reasonable return of telephone calls
- Add-on value for customers (over basic service)

*Primarily subjective* quality attributes

- Openness
- Convenience
- Responsiveness
- Helpfulness and usefulness
- Good value for money
- Healthy and pleasant environment
- Courtesy toward clients, users
- Respectfulness
- Interest in providing values sought by users
- Committed to improvement

The questions following each chapter are typical questions often asked about the topic, or are included here to draw the reader's attention to some important point. The suggested answers are provided as food for thought. The suggested answers have the same numbers as the corresponding questions.

# QUESTIONS

Q: 1-1    Why is productivity improvement vital?

Q: 1-2    How large of annual increases can be achieved in productivity?

Q: 1-3    What are the two main types of productivity improvements everyone should try to achieve?

Q: 1-4    Who benefits from improved productivity?

Q: 1-5    What is the difference between productivity and profitability?

Q: 1-6    Does higher quality mean lower productivity?

Q: 1-7    Do higher sales mean higher productivity?

Q: 1-8    What do "inputs" and "outputs" mean?

# SUGGESTED ANSWERS

A: 1-1    Productivity is essential today because the consumer is now better educated, is more demanding, and wants value for money. Furthermore, greater global as well as domestic competition demand more and better goods and services at lower prices. Governments are also forced to produce and provide more from less. Workers are demanding stable employment, higher pay, and better quality of working life. None of these demands can be met without increased productivity.

A: 1-2    Evidence shows that up to five percent annual productivity increases or even more are realistic for organizations.

A: 1-3    There are two main aspects of the necessary productivity improvements: improvements in quantity and in quality, that is, produce more and/or better from the same resources.

A: 1-4   Everyone benefits from improved productivity, for example:

- Employers: through greater competitiveness, stability, and higher profits

- Employees: through more secure employment, better wages and working conditions

- Consumers: though higher quality and lower prices

- General public: through greater prosperity and relatively lower taxes

A: 1-5   Productivity is the relationship between the volume of output and the volume of resources used. For example, the number of chairs produced per worker-day. Profitability, on the other hand, is a value relationship between the value of production and the costs of production. In the long run, profitability is determined largely by productivity, but in the short run, price fluctuations and variations may hide the relationship. In practice, it is useful to measure and analyze both productivity and profitability, possibly together, in an integrated framework.

A: 1-6   Quality and productivity usually support each other. For example, defect-free production means higher quality but also higher productivity because of eliminating waste, rework, re-administration, warranty costs, and so on.

A: 1-7   No, greater sales do not necessarily mean higher productivity, because higher sales may result from the use of more resources or inflationary price changes. Let us remember that productivity is a relative concept: output per input.

A: 1-8   In the commonly used, *efficiency*-type productivity measures, inputs are the organization's labor and capital, plus whatever is purchased from others and used in producing the output or providing the services, such as materials, components, fuel, and energy, as well as services, including general consulting, accounting, and legal services. These inputs are used to produce or provide the organization's *output,* that is, the organization's production or services. In *effectiveness*-type productivity measures—which are sometimes used, mainly in public and noncommercial services—the same type of inputs are related to the *outcome,* that is, the results achieved by the organization's *output*s that had been provided by the resources used, such as fewer highway accidents with the same police protection.

# 2

# Special Productivity Improvement Needs of Smaller Enterprises and Services, Including Professionals

## WHY IS PRODUCTIVITY VITAL TO SMALL AND MEDIUM-SIZE ENTERPRISES?

Small businesses, or rather small and medium-size enterprises, abbreviated SMEs, are usually not little "big businesses." They have their own reasons to exist, as well as their own characteristics. There is a great variety of SMEs. Their differences need to be taken into consideration when one explores why productivity improvement is vital for the survival and prosperity of small businesses, and how their productivity, profitability, and competitiveness can be enhanced.

An SME is basically a goods-producing or service-providing enterprise that is more or less independently owned and operated and is not dominant in its field. Some may employ just a few people while others have many employees, even up to 500, but are still considered "small" in their industry. Relatively large "small" manufacturers have been recorded in such industry sectors as the manufacturers of aluminum windows and doors and wooden windows and doors, steel foundries, steel fabricators, hosiery manufacturers, their retailers, and others.

Smaller firms vary from one another in their purpose. Many have been created by their owners to make their own product, make a better use of their skills, meet the challenge of new opportunities, make a better product, provide a better service and, at the same time, earn a profit. Some want to use their own approach to their work or be their own boss. SMEs can be small manufacturers or such businesses as technicians, mechanics, retail storekeepers, and restaurant or marina operators. Others want to be personal or business service entrepreneurs, including sports trainers, barbers, hairdressers, or travel service operators. Some smaller firms are very successful even in the high-tech sector, for instance in computer services, because they offer specialist services or high-quality technical support to their large-scale competitors. These companies may eventually grow into large companies or develop into chains of SMEs. Other small companies are long-established organizations that are still small because of their nature of providing a service to the population of a relatively small area, or are smaller manufacturers who supply a local or even worldwide niche market.

While SMEs are widely different, there are nevertheless certain characteristics that tend to be common to most. For example, smaller firms have social and environmental advantages, are more flexible, can move quickly, and have the ability to meet the demand of limited markets. SMEs can have firsthand knowledge of the customers' needs and can adjust to make their products or services meet these needs. This ability gives smaller firms a competitive edge. In smaller firms, the owner/manager looks after the whole business while large corporations need to have lots of checks and balances among the various functional departments. The large organization requires a great deal of communication and the preparation of a large volume of reports, many of which deal with small parts of the business. This represents much of the costs of large corporations. The capital intensity of SMEs is typically lower than that of large firms, although the productivity of the smaller enterprise is not necessarily lower at all. SMEs can make a profit on limited production volumes that would not be profitable for large enterprises because of their relatively high overhead costs.

In view of the large variations in company scale, it is natural to ask whether various company characteristics are related to scale. We have undertaken a variety of analyses and have found that in most industry sectors there is no basis for assuming that successively larger companies benefit from scale increases. Our observations have also proved that there is a tremendous variation in productivity performance between companies in the same size group, as well as between all companies in the same sector. On the other hand, various studies have shown that product-specific scale increases do exist, that is, productivity benefits can be derived from increased lengths of production runs of similar products in plants of any size.

The mistaken assumptions of generally valid company economies of scale may have originated from macroeconomic data correlating, for example, company size and productivity for manufacturing as a whole. We have found strong indications that this resulted from the fact that less-productive industry sectors tend to be smaller and the more-productive industry sectors are larger, hiding the fact that within individual industry sectors there is in most cases little evidence, if any, of a relationship between company productivity and size.

The disadvantages of SMEs tend to be more specific. For example, they may demand more attention to productivity matters on the part of the entrepreneur, who is very often a technical person or specialist rather than a managerial specialist. The personality and ability of the individual manager or of the head of a small management group play a very important part in determining the success or failure of the business. They have to prepare plans and make a variety of decisions often based on a hunch rather than thorough analysis. As will be discussed later, some productivity measures and analyses, even relatively simple ones, tend to pay off well, and most of the information in this book should help managers a great deal.

Our experience around the world shows that the prosperity and often even the survival of SMEs are largely determined by their productivity. Our data have confirmed and reconfirmed that practically all of the analyzed SMEs have a very large potential for improving their productivity, up to 80 percent between the median and best comparable firm. In order to reach higher productivity it is essential, however, that managers of SMEs understand the meaning and importance of productivity. Smaller entrepreneurs usually respond to challenges that they understand.

Yet, many small business managers are unaware of the existing problems and opportunities, and many do not know how productivity, quality, competitiveness, and profitability can be increased substantially. The need for understanding productivity and being conscious of its vital importance is particularly great among such enterprises because their managers, often owner/managers, do not have all kinds of professional specialists at their side, and must make technical, managerial, organizational, as well as financial and myriads of other decisions by themselves. One of the main purposes of this book is to help entrepreneurs and their staffs help themselves to improve their own productivity and that of their organizations. This chapter intends to draw attention to the special needs of smaller enterprises.

The key to improving productivity is diagnosis of the organization's performance and productivity. Managers and entrepreneurs need meaningful data and at least simple analytical methods to be able to see for themselves how they are doing and where they could improve. Managers can usually benefit from learning how to cooperate with their peers, employees, suppliers, and customers, as well as other companies within and outside their industry.

Many benefit from *benchmarking*, that is, from comparing one business's management methods with other successful enterprises. Usually this comparison will not be with direct competitors. One can learn a great deal, including new methods, from companies who have succeeded to excel in certain functions, say, administration, personal relations, design, production, warehousing, or distribution. Even leading large companies have benefited from this learning method. If mutually arranged interfirm comparisons can be made between comparable and competing companies, it is even better!

One also can learn a great deal from consultants, who can help find new opportunities or identify the causes of problems. Managers are often surprised to find that the problem they are facing has already been faced by many other firms and that various relevant solutions have already been found and tested. While professional consultants are relatively expensive, it has often been proven that the service they can provide for SMEs tends to pay off. Much managerial advice can also be obtained from specialists who are not consultants, such as bank managers, accountants, industry or trade association leaders, and so on. Topics that SME managers need to know include the methods of financing, planning, scheduling, and controlling work, handling personnel matters, evaluating performance, managing inventory, marketing, and collecting receivables. The skills needed—but often missed—by SME managers include methods of keeping simple but reliable production and cost records and basic productivity and financial ratio systems, which will be discussed later on.

# MEASUREMENT OF RESULTS IS A MUST FOR SME MANAGERS

The potential for productivity improvement and the resultant higher profitability is there for the smaller company. A conscious effort must be made, however, to realize it. Good decisions must be based on solid analyses. One must know the facts before one can take effective action. The weaknesses of the past must be identified before they can be corrected, and the opportunities must be known before they can be exploited. A very conscious effort must be made to undertake a systematic and comprehensive—even if simple—analysis in which the various interrelationships are revealed. All problems must be approached with an open mind because the so-called "conventional wisdoms" are often not so wise or even applicable in particular areas.

The various requirements and methods of productivity measurement will be discussed in detail later in this book, but some special considerations relevant to SMEs are brought to your attention here. The demise of many small enterprises was, at least partly, due to inaccurate, incomplete, and inconsistent cost and output accounting information available to their executives. The data should show the manager or entrepreneur which of the output and input areas need to be investigated with somewhat greater precision or urgency. An integrated diagnostic appraisal is new to many SMEs, but it is very important to look at a small firm as a whole. The analysis may focus on financial appraisal but must go beyond the financial data, which may hide the underlying efficiency and capacity utilization factors and which can be biased by external financial factors that are beyond the firm's control, such as inflation. One often hears arguments that this or that function can not be measured but, in fact, most functions *can* be measured if they are clearly defined. Many aspects of how this can be done will be discussed below in great detail. A few examples should suffice here.

Although SME managers tend to detest the paperwork or computer files necessary to keep good records, such data are essential for them because SMEs tend to be much more vulnerable than larger competitors to economic fluctuations and external forces, as well as such matters as liquidity problems in cases of rapid growth. It is important that the information be straightforward, clear, and not overwhelming. SME managers must also know how to make use of the information gathered. Even if they have difficulties in the beginning, it has been our finding that as time goes by, SMEs working to improve their productivity make gradually more and more use of the information and data, including integrated productivity and cost ratios. In the past, the data available in smaller firms often lacked meaningful productivity measures, as well as meaningful breakdowns of their production, marketing, and other major costs. For example, a business with, say, 10 to 20 products or services should not look only at total sales but find out which products are less profitable or not moving quickly enough and try to adjust their product mix.

Productivity measurement and analysis have consistently proved to be effective tools of self-education for smaller enterprises because they induce the entrepreneurs and managers to become systematically and fully familiar with their production and service processes, together with the related financing, managerial, personnel, as well as marketing and distribution aspects. As a result of such analyses, smaller firms have often enhanced their performance by improving the planning and forecasting of such variables as sales, inventories, direct and indirect labor requirements, and the utilization of plant, energy, water, and consumable materials. Need for improvement was also often indicated by the analysis of warehousing, distribution, or traffic routing. Any foreseeable bottlenecks could be usefully identified and avoided. Low productivity has also been caused by inadequate or inappropriate education or training of staff, or by low wage levels that cause excessive labor turnover or overly high training costs. SME managers have to recognize these defects and provide for the necessary training or adjustments.

It is, of course, important to relate the measures between themselves. For example, a company selling for the general market is bound to have higher inventories than another firm that produces on a custom-order basis. Also, a company with high capital intensity should normally be expected to show high labor productivity, while a company having unutilized capacity might benefit from increasing its sales by adjusting its pricing or advertising policies. In closing this section, it should be emphasized again that it is important to keep the measures simple, choosing only the meaningful ones on which you can act, and not giving in to a seemingly desirable proliferating of unnecessary information or detail.

SME managers should keep in mind that there are two kinds of productivity analyses that can be usefully carried out, namely *productivity trend analyses* over time and *cross-sectional analyses,* which compare various SMEs at one point in time. Both are useful for their own purposes, and their combination can be even more revealing. In many instances, SMEs must be satisfied with less than pure productivity figures due to data limitations. Productivity analysis can be very effectively combined with profitability analysis, as our interfirm comparisons have shown. Managers of thousands of companies that took part in our studies were amazed how much better insight they gained into their own company than they had before the comparative analyses.

Productivity analysis and productivity improvement go hand in hand. Nevertheless, at the present time very few smaller firms have an effective system for measuring variations and changes in productivity relationships, and even fewer know the causes of such variations and changes. The analyses should show the managers/entrepreneurs which of the output or input areas need to be investigated in more detail or with a somewhat greater urgency. A company manager needs to know whether his results are achieved with the optimal combination of human, physical, technical, and economic resources. He should know his fixed and variable costs and the underlying productivity relationships so that his prices will accurately reflect his costs and indicate when relative price and cost changes may make it desirable to shift to relatively cheaper resources.

# STEPS TO IMPROVE THE PRODUCTIVITY OF SMEs

Our studies of SMEs have shown that after the companies had identified their weaknesses, the corrective actions ranged throughout the whole spectrum of business operations. In the area of output of manufacturing enterprises, a variety of corrective measures have been reported, including increases in shop size to allow greater sales volume with little increase in overhead. In some cases, the specified objective was accomplished by changing the product mix in order to increase average shop labor productivity and reduce the corresponding labor cost. The reduction of product variety, numbers of models and procedures, as well as changes in output quality or price structure were among other corrective steps reported in the output area.

As labor costs are a major element of total costs in most industry sectors, many of the analyzed companies have given labor productivity improvement a high profile. Some firms have introduced a continuing audit of labor productivity and efficiency, including the monitoring of the number of employees in relation to goods produced. Companies with high employee turnover have initiated action to identify its causes and eliminate them. Others improved manpower utilization through supervisor training, better management control, or improved motivation.

Capital productivity, that is, the efficient and effective utilization of plant and equipment, has received considerable attention in many manufacturing SMEs that participated in our interfirm productivity/profitability comparisons. Some have taken measures to improve the utilization of their fixed assets, while others have taken action to improve their machinery or increase their mechanization. Many firms have taken steps to increase their automation while others have decided to sell some unproductive assets. In the field of current assets, the analytical information has induced a large number of SMEs to take steps to improve the turnover of their working capital by adjusting their inventories, revising the company's billing and collection systems, and tightening the control of accounts receivable.

In service industry SMEs, for example, hotels, motels, and resort establishments, the major productivity improvement opportunities have been observed in the fields of management, finance and accounting, manpower use, and marketing. In these establishments, managers need to learn and utilize modern management skills and improve the use of their time. They need to set solid goals and design realistic plans. In order to achieve these objectives, they need to measure and evaluate their operation. Managers must ensure that the cash flow available is adequate to compensate the owners/managers for their work and achieve satisfactory return on their investment. In the area of finance and accounting, hotel managers need to establish financial objectives, determine operation targets, as well as prepare and use at least monthly operating reports. Marketing often needs strengthening, particularly in pricing, in-house advertising, as well as regular and special promotion activities. As managers of SMEs often need to be absent or otherwise unavailable, an operations manual needs to be prepared so that their substitutes can always know how the business operates. There is a need to attract, orient, train, and retain quality staff, and set satisfactory wages, fringe benefits, and incentives. In order to optimize the use of staff, hotels and other similar establishments should train their employees to various functions.

In the rooms department, regular record keeping and inspection is necessary. We have found comparisons with other similar establishments to be particularly useful as the productivity of room personnel has shown very large and unjustified variations, such as in the number of rooms looked after by a maid. For example, at some resort hotels—where walking distances tend to be great—some managers thought that 10 rooms per maid was satisfactory until they learned that in some of their competitors each maid could do 12 rooms per day. It is, of course, different in high-rise hotels, where the maids can look after 16 to 18 rooms. In the foods department, managers should attempt to increase sales volume and improve quality, including service skills. All costs, including purchase costs, need to be watched and minimized. Incoming ingredients need to be inspected, portion size and quality controlled, and waste avoided.

In the beverage department, the sales/costs relations need to be optimized, high quality ensured in mixing drinks, and inventory should be minimized by keeping the most popular drinks plus only a few premium brands. Sales and inventory in bottles need to be monitored. The miscellaneous department also should contribute to overall revenues and be monitored.

# PRODUCTIVITY NEEDS OF PROFESSIONALS AND OTHERS IN SERVICES

It can not be emphasized and reemphasized often enough that productivity and overall success in the service sector, including professionals and other knowledge workers, can be greatly improved. The need for this has been observed in the offices of doctors, consultants, engineers, the information technology (IT) sector, banks and other financial institutions, as well as in the public sector.

Their training in management and productivity often tends to be weak. This is also true for service occupations in all physical goods–producing industries including manufacturing and the resource industries. Productivity is vital in services in order to satisfy the growing demands of customers, clients, and patients, to fight increasing competition with fewer human, physical, and financial resources, and to take advantage of opportunities offered by deregulation, as well as rapidly increasing information technology and the internationalization of many services.

The improvement is far from automatic. It is essential that all people in service work, including professionals, other knowledge workers, and indirect workers, understand what productivity is and how they can improve it. Their training is heavily focused on their specific expertise while their general managerial knowledge and special productivity expertise—which go hand in hand—can in most cases be enhanced a great deal. Studies have shown that the real return on productivity investment in this field is very high.

By now, people in service occupations form the major part of the workforce. Even in manufacturing and other goods-producing industries the number of service functions is rapidly increasing. Progress in science, and rapidly increasing mechanization, automation, electronics, and information technology enhance the need for engineering, control, maintenance, accounting, and clerical work. It is in everyone's interest in all these fields to make the most effective and efficient use of all of their resources.

The proportion of knowledge workers and indirect workers, that is, those workers whose work can not be identified with any specific product, is growing because:

- The rapidly growing sophistication and complexity of both products and services demands a much better educated and informed labor force

- The rapid integration of global markets and global competition requires better trained workers

- Intensifying technological progress enables much more work to be done by complex machines, which require better trained operators

- The general economic advance and the increasing proportion of service jobs

In order to understand the special productivity needs of professional and other knowledge workers, as well as of all indirect workers, it is necessary to examine who belongs to these groups. The categories are not always clearly demarcated and often overlap, but an attempt to distinguish them helps to understand their relative productivity needs.

The basic distinction is between knowledge workers and indirect workers as distinct from skill-based workers. The tasks of knowledge workers depend on know-how, tend to be creative, require judgment, and are often developmental in nature. The knowledge worker group includes all professionals, such as engineers, doctors, nurses, teachers, economists, lawyers, accountants, and highly trained IT workers, as well as electro-technicians, X-ray operators, laboratory technicians, and bookkeepers.

Other knowledge workers are indirect knowledge workers, for instance those who work in office tasks of management, personnel, administration, finance, and accounting, computer programming, design, quality control, banking, sales, and cashier, clerical, or secretarial occupations. Not so long ago, many of these workers were known as "white-collar" workers. Today this distinction from "blue-collar" workers has largely disappeared. Just to mention one example, we visited a factory that manufactures transmissions for large trucks and buses, and all the "production workers" wore white shirts.

Not all indirect workers are knowledge workers. Their work may be, say, in the manufacture and distribution of products but can not be allocated directly to a specific product or service. It does not affect the size, shape, or condition of a specific product. Examples of relevant occupations are in receiving, maintenance, inspection, materials handling, warehousing, shipping, security, and janitorial work. Different from professionals, other knowledge workers, and other indirect workers are the skill-based workers who follow clearly defined procedures, even if these may be difficult, complex, and require long periods of training. This group includes, for example, carpenters, locksmiths, and tool and die makers.

All can—and should—learn to understand their need for productivity improvement. Their specific needs vary, however, from group to group. Some can benefit more in one area than others. Most can gain from learning how to plan their work, how to measure and analyze their own productivity, and how they can benefit from the use of information technology. They all benefit from learning that they need to keep updating their knowledge, including that of understanding and measuring productivity.

However, their further needs vary from one occupation to another. Professionals, for instance, need to learn more about how to forecast, organize, and balance their work, manage staff and resources, motivate others, and communicate in an effective, courteous, and pleasant manner. For example, professors in rapidly advancing fields, such as biochemistry, can often improve their own productivity and that of their staff by better forecasting, planning, and organizing, accepting only responsibilities and assignments that they can meet in a timely fashion, better utilizing their time, and balancing the workload of their staff. For other knowledge workers and most other indirect workers, setting priorities, managing their time, and communicating with others are areas where they can particularly enhance their productivity.

Skill-based workers can also improve their productivity significantly by, among other methods, updating their particular technical skills and organizational knowledge, as well as by making full use of their time and space. Salespeople have benefited from systematically analyzing their target market, better focusing on promising customers, as well as improving their negotiating and interpersonal skills. In airlines, considerable productivity and service improvements of staff have been made by improved communication, such as regular discussions of problems and solutions among flight staff, for example, chief stewards and flight attendants, both between themselves and with the flying public.

# MEASUREMENT NEEDS OF THE KNOWLEDGE GROUP

Professional and other knowledge workers need productivity measures to spot weaknesses as well as opportunities, identify efficient and inefficient areas of work, evaluate programs, assign priorities, and justify resource requirements. Measures can show, for example, how many employees they have compared to how many they need.

Even in manufacturing it has been estimated that without measurement and related control, indirect labor performance does not run much above 50 percent of capacity. For example, much of the "attendance" may not be "on the task." It is, therefore, often desirable to observe the distribution of one's time spent on various functions. For example, in one observation, a quarter of supervisors' time was spent on writing, one-third on talking about technical aspects of work, 10 percent on personal matters, and so on. For nonsupervisors, about 40 percent of the time was used for setting up experiments and taking readings and one quarter of the time for talking about the technical work.

The productivity measures of professionals, other knowledge workers, and other indirect workers need to be simple, meaningful, and usable. It is of limited use, if any, to focus on measurement approaches relying on meaningless statistical counts, for example, number of publications, reprints, or citations, because the quality and significance of the various contributions vary a great deal. Measures for all knowledge workers must have a particularly high degree of acceptance and support at all levels of workers because the persons in these groups tend to work more independently, "on their own." We also have to keep in mind that in R&D there may be a long delay between the relevant inputs and outputs or outcomes. If total output or outcome is difficult to measure, it may be useful to differentiate between phases, activities, or tasks.

For this group of workers, it needs to be kept in mind that formal evaluation, particularly quantitative evaluation of individual output, can easily discourage employees and may be counterproductive. It may bias production in order to best fit the measures applied. In some cases, therefore, it may be preferable to look at the productivity of the organizational unit rather than individual productivity. (Members of the group will probably ensure that all participants will do their share.) For R&D, value analysis is among the recommended measurement methods, concentrating on various aspects of relevance of the output to needs, such as urgency or payoff, for example, who is waiting for the results or what are its leverage effects. Other recommended measures include institutional health, such as the probability of success of a research experiment, technical merit, staff welfare, and so on.

Special attention must be paid to the quality and timeliness of performance in these types of work. Comparisons can be made with forecasts and norms. It must be ensured that only comparable outputs and inputs are compared. Actual output costs can be compared with standard costs. Production may be converted into standard person-hours and compared with actual person-hours. This can be taken as an indicator of productivity.

The inputs are usually clear and easily measurable, even for R&D. It is useful to classify labor inputs into direct personal services, overhead services, and other categories. Measures of indirect work often include ratios of number of units of output per length of worker time, number of error-free bills prepared per employee-hour, number of drawings prepared per employee time, number of inquiries answered satisfactorily, or the number of products shipped in error. Nevertheless, in every case, the data need to be realistic and relevant. In some cases, the input of indirect labor has also been measured by "deflated" salaries in order to give weight to the quality and qualifications of labor.

# EXAMPLES OF PRODUCTIVITY IMPROVEMENT IDEAS FOR PROFESSIONALS, OTHER KNOWLEDGE WORKERS, AND INDIRECT WORKERS

The productivity of professionals and other knowledge workers can be greatly improved if they consider themselves as a business that has human and physical resources at its disposal, which need to be utilized effectively and efficiently. Service orientation, strategic direction, as well as modern management style and techniques are of primary importance. Effective work methods need to be chosen in order to simplify work and subdivide it into better manageable units. Motivation of staff, setting milestones, and completion dates for projects are needed. Workloads need to be balanced to avoid staff friction.

The productivity and quality of professionals, knowledge workers, and most other indirect workers can be improved by:

- Developing their productivity and quality awareness.

- Planning carefully.

- Defining the task clearly.

- Determining whether the job is really useful and needed, by whom, and why. For example, "busywork" that adds little to what they are paid for should be minimized.

- Utilizing all personnel and physical resources.

- Improving the timeliness of the work; responding rapidly to requests.

- Simplifying and standardizing actions where feasible.

- Communicating effectively with clients, customers, and patients, and among the various levels of staff.

- Keeping good records and ensuring that one's staff also keep adequate records in order to avoid the need for repeating lost work.

- Measuring productivity performance and providing feedback to the persons involved.

In many professional jobs, such as in research, diagnoses of physicians, policy work, strategic decision making, and editorial work, quality is a much more important determinant of productivity than quantity.

Among examples of methods of increasing the productivity of office work are:

- Focusing on the mission of the organization: "what we are trying to achieve"

- Listing jobs in writing and prioritizing them with regard to the overall objectives:

  - Eliminating unnecessary duplications, redundancies, backtracking

  - Ensuring accuracy, clarity, and simplicity

  - Making the best use of time and other resources

  - Determining the needs of the clients and focusing on them

  - Ensuring accessibility, courtesy, friendliness, and helpfulness

In affiliates or branches of larger enterprises, productivity/profitability monitoring enables the comparison of methods and routines of individual affiliates or branches and leads to identifying the causes of variations in the results. For example, productivity in banks can be improved by:

- Determining what the various customer categories want and are willing to pay for

- Providing the right products at the right price

- Developing the required products in standardized forms

- Concentrating on the "cash cows" and providing the "loss leaders" with satisfactory and suitable but simplified and standardized products, as well as by introducing self-service if possible

- Adjusting business hours to customer requirements

- Making services available by Internet, telephone, fax, or correspondence, and utilizing economies of scale as much as possible

Another example of the importance of productivity in services is that of police service where studies have indicated the value of cutting the requirement of preparing unnecessary reports, reducing waiting time in courts, allocating resources to services of highest value, anticipating the occurrence of crime by location and time, measuring and rewarding human resource performance, as well as making the most of talent and physical resources. Among productivity improvements reported by police in the use of physical capital were the enabling of quick turnaround of their motor vehicles from service shops, for instance, by the introduction of removable communications equipment. When a computer in a car breaks down, the computer is immediately replaced by a working piece of equipment so that the car is not sitting idle while the communications equipment is being repaired.

Among other service fields, firefighting services can learn a great deal about their productivity opportunities and weaknesses by measuring, recording, and analyzing how and where their time is spent, identifying idle resources, and prioritizing according to what the public expects. Firefighting services have also reported that using better tools, such as circular saws instead of axes, has improved their productivity tremendously, as has combining the firefighting service with emergency ambulance service since most firefighting services have surplus capacity.

For retail stores, vital productivity maximization is achieved through making the best use of staff, space, and inventories and offering highly sought products at competitive prices, as well as ensuring better customer and office services. In many service jobs, productivity improvement can be achieved by increasing the quantity of output by improving organization. Such jobs include the filing of documents, handling insurance claims, doing nursing assistant work, and so on.

Later in this book many of these issues will be discussed in greater detail.

# QUESTIONS

Q: 2-1   Are SMEs not competitive with larger companies because of their size?

Q: 2-2   Why is productivity training very important for SME managers?

Q: 2-3   What is benchmarking?

Q: 2-4   Why is productivity improvement essential in hotels?

Q: 2-5   How can one increase productivity and profitability in service operations?

Q: 2-6   Why is the understanding and improvement of productivity particularly urgent for professionals and other knowledge workers?

Q: 2-7   How can professionals and other knowledge workers benefit from planning?

Q: 2-8   How can professionals and knowledge workers benefit from a clear desk?

# SUGGESTED ANSWERS

A: 2-1   SMEs can be competitive with larger companies because their productivity may be enhanced by their flexibility and ability to have shorter production runs due to specialization and niche marketing.

A: 2-2   Productivity and managerial training in general is especially urgent for SME managers because they are often entrepreneurs with a technical or nonmanagerial background and usually don't have managerial specialists on their staff.

A: 2-3    Benchmarking is a comparison with and learning from other companies—usually not competitors—that have known success and expertise in some aspect of business, such as planning, scheduling, personnel management, or physical distribution.

A: 2-4    Hotels, motels, and resorts need productivity improvement because it reduces costs in operations that are the main determinant of room and restaurant rates, it enhances quality, and, as a result, attracts, satisfies, and retains guests.

A: 2-5    Some examples of what can be done in order to improve productivity/profitability in services are as follows:

- Service providers must determine what the clients/consumers want and are willing to pay for.

- Differentiate the services to meet the needs of segmented groups of consumers.

- Shift the sales toward the more profitable clients, and standardize these services.

- Steer customers toward the more profitable services.

- Apply all the best methods of time management and resource utilization.

A: 2-6    Productivity improvement is particularly urgent for professionals and other knowledge workers because their training is focused on their specialist expertise and not managerial and related productivity training.

A: 2-7    Planning saves wasting time on unnecessary and repetitive actions and uses the available resources to the best advantage.

A: 2-8    Professionals and knowledge workers can benefit from a clear desk because an uncluttered desk is essential for proper organization and prioritization of activities, and it enables efficient work on the task at hand.

# Part II

## How to Identify Productivity Problems and Opportunities: Measurement and Analysis

# 3

# You Need to Know the Facts before Taking Corrective Action

## OBJECTIVES AND USES OF PRODUCTIVITY AND QUALITY INDICATORS

The prerequisite and first step of improving productivity and quality is a thorough analysis of the underlying factors. Unless we can identify the problems through systematic and adequate quantitative and qualitative diagnosis, it is not possible to find effective corrective actions. "Any" medicine is not better than no medicine. Unsubstantiated guesses or improper analysis may result in a "cure" that could do more harm than good. Good analysis, on the other hand, does not only reveal problems but also opportunities for fruitful actions.

Productivity measures do not replace conventional financial accounting but complement it and substantially strengthen it. A critical advantage of productivity measures for performance analysis is that they do not look at costs alone but quantify the resource utilization. Conventional accounting information typically does not provide all the necessary information for productivity and quality measurement—neither soon enough for productivity improvement action nor in sufficient detail—and does not relate outputs to their inputs in real terms.

Furthermore, traditional financial accounts are not suited for productivity measurement because they are:

- Expressed in current dollars, subject to changes in currency values

- Affected by factors irrelevant for productivity measurement, for example, overtime hours paid rather than actual hours worked

- Developed for taxation purposes according to taxation rules, which often do not yield realistic data needed for productivity measurement

- Short of various specific data that are needed for productivity and quality purposes but are not gathered for financial accounts

Although many data that are needed for productivity measurement are already available from conventional accounts, they must be adjusted for productivity measurement purposes. For example, these data need to be converted into real terms and supplemented by other data required for productivity measurement, for example, person-hours worked rather than person-hours paid.

Proper analysis must be based on facts, solid quantitative information. We must find out where we are in terms of performance. What and how much do we produce, and what resources and how much of the resources do we use to provide the goods and services wanted? Where is it the most important to use our available resources? Where are the gaps between our performance and the customers' or clients' expectations? In each of these terms, how do we compare with other similar organizations and organizational units? Are we doing better or worse than others? By how much, and why? Or, how do we perform compared with the past?

The analytical procedures should meet the objective of improving systematic decision making, both in the determination of the organization's overall policies and the internal management of the establishment. Input and output measurement is equally vital in both the private and public sectors. From Westinghouse to IBM, from Xerox to Hitachi, successful corporations base their productive and competitive performance on reliable measurement. Quantitative performance information is of equally critical importance to smaller and medium-size establishments. In the public sector the measurement of productivity and quality ensures that public funds are spent efficiently and effectively, with optimal results.

An example from the physical distribution industry can illustrate some of the above points. Data on loading and unloading weight per labor hour or equipment hour and ton/kilometers per hour, for example, help determine the required size of the fleet, enable control of the operating performance, and indicate the effectiveness of the utilization of warehouse crew and capacity. Measures showing the savings in labor hours per dollar of investment in equipment help the calculation of the trade-off ratios between the respective inputs of labor and capital, as well as the justification and proper choice of capital investment.

Some of the major specific functions of management in which productivity measures are particularly useful are as follows.

## Goal Setting and Planning

Productivity measures help in goal setting because they force the clarification and better understanding of the goals and the underlying assumptions, taking into account inputs and their costs as well as the expected results. They help create productivity consciousness in the goal-setting process. The productivity measures stimulate the search for completeness in the specification of goals, help to group similar subobjectives in order to eliminate unnecessary duplications, and guide the decision makers toward planning subobjectives in order of importance.

The efficiency indexes indicate changes in the real cost of producing the organization's projected output. The effectiveness measures reflect the value of the organization's output of goods and services to the customer. An analysis of past productivity levels and trends helps in making more efficient and effective decisions for the future because past measures show what happened to productivity under various circumstances.

The major components of total output and input, as well as their respective proportions within the totals, should be analyzed separately. When financial data are related to output measures, they can show their current relationships as well as the changes in these relationships over time on a per unit basis. This analysis makes it possible to break down total unit costs by type of cost. The partial cost ratios, in turn, can be analyzed in comparison with the relation of the costs to the total output. For example, direct production labor cost/sales value of production as compared with total production cost of sales/sales. See more examples in Chapter 7.

The detailed measures enable the planners to project their resource needs more accurately. The data help them answer such questions as "How many people are needed for this job?" or "How much capital is needed for this work?" The data help in setting the relevant wage and merit pay policies, as well as personnel policies regarding such matters as hiring, firing, transfers, promotions, and rewards. The analytical process shows where the existing productivity indicators are insufficient for management purposes, for goal setting, as well as for performance monitoring, control, and evaluation.

The goals can be assessed in terms of geographic variations of productivity and their causes. The overall effectiveness of the system can then be improved, say, by shifting resources from one subobjective to another. For example, in the health care area it has been found that the consumption of medical care varies greatly by geographic area. If the reason for such variation can be determined, overall productivity can be increased in the provision of medical services, say, by shifting resources from curative care to preventive care where this is indicated by the regional productivity data.

## Budget Justification

Productivity measures also help in the budget process. They reveal the financial implications of alternatives. They enable a clear identification of the relationship between the operational plan and the budget. Clear reasoning and calculating resource requirements ensure that the reviewers of the projected budget will not look at the proposals in an arbitrary way. It is also important to make productivity indicators an integral part of the budget process because managers will pay attention to productivity measures only if these are incorporated in the budget formulation and review process.

## Performance Improvement

Productivity measures help identify the best course of action in terms of resource utilization and cost-effectiveness. They help to improve scheduling and to plan the best location and arrangement of the physical facilities. They can show what factors influence the results the most so that emphasis can be placed on those. Productivity measures raise questions of good or bad performance that require explanation. They help identify the reasons for outstanding performance from which others can learn. The data can spot and improve weak or lagging areas and lead to reduced resource requirements to achieve the previous level of output or, alternatively, to increased or improved output with the same resources.

Productivity measures can explain variations from budget, why they occur, and what is responsible for them. They identify uneconomic activities, wastage, or underutilized capacity. They can find out which factors can be manipulated in order to increase productivity and profitability. Procedures of questionable efficiency can be identified and replaced by more efficient ones.

Productivity measures also help in avoiding misinvestments. The introduction of a high-tech piece of equipment is attractive to managers. Experience indicates, however, that it is essential first to evaluate the productivity and cost implications of the introduction of the planned investment, say, of new computers because there is the danger of going into heavy expenditures without positive results. First, it is necessary to measure and analyze the productivity of the operation in order to eliminate the inefficiency of the present system.

Productivity measures can indicate areas where actions can be taken to achieve a certain objective, and whether the actions taken have led to the intended improvements. The impact of various actions, such as changes to organization or procedures, adding new equipment, or introducing new rules or legislation, can be judged better. Last, but not least, productivity measures can help in the proper allocation of central costs (for example, that of financing, accounting, personnel, or legal activities) to the various organizational units.

## Control of Operations and Accountability

The creation of a productivity measurement system tends to clarify the stages of program performance and assist in the evaluation of actions taken, including that of organizational changes. A productivity index will reflect departures from the past trend, from set objectives, or from the performance of comparable organizations and provide a solid basis for making improvements or taking corrective actions.

Clearly communicating the desired level of performance and the actual performance tends to motivate people to improve productivity. The introduction of effective managerial accountability is also made possible by productivity measures because they reflect the real relationships between the products or services produced and the resources used. The productivity data form an objective basis for the determination and rewarding of good performance, as well as for allowing improvement of poor performance.

## Productivity Analysis As a Training Tool

Experience has proved that the productivity measurement exercise is, in itself, an excellent productivity training tool and, in general, an outstanding management training method because it brings together in a very clear and systematic framework what is produced and the resources used in its production. It enforces the clarification of all inputs and outputs as well as their relationship, and leads management to a thorough familiarization with the organization's objectives and processes. Management learns exactly what is being done and why, and what resources are consumed in production and service.

Finally, productivity measures are also very useful public relations tools because they provide factual evidence, in very specific terms, of the performance of both private- and public-sector organizations to the media and the public, helping to justify budget requirements.

## General Approaches to Productivity Measurement and Analysis

The objective of all productivity measures is to identify productivity weaknesses and opportunities and thereby enable productivity improvement. The form and detail required will vary from case to case but it should always be kept in mind that the collection, adjustment, and calculation of data costs money and should be kept to the necessary minimum. The planners must always ask themselves such questions as: "What is this information needed for?" "Who will need it?" "Why is it needed?" "What will it show?" "What question will it answer?" "Can follow-up action be taken?" and "Who will take the corrective steps?"

The best way to start is to try to account for 100 percent of the time taken by the activities of the unit (or individual) whose productivity is to be measured. This can be done in, say, five or 10 percent intervals. For example, 60 percent of the time may be spent on treating patients, 10 percent on keeping records, 10 percent on keeping up to date with developments, five percent on discussion with colleagues, five percent on obtaining supplies, five percent on meetings, and five percent on other things.

In another example, what percentage of the supervisors' time is spent on organizing production, setting standards, attending meetings, and looking after personnel administration, tools, equipment, materials, and other activities? What percentage of their time is spent by their staff working alone, with supervisors, with subordinates, with staff, or others? It is also useful to estimate the percentage distribution of the time personnel spend on various activities, such as handling, walking, looking, showing, talking, listening, standing, sitting, reading, writing, calculating, or other activities. In technological types of work, the supervisors may spend as much as a quarter of their time writing and perhaps even more talking about technical work. In laboratories, nonsupervisory personnel may spend almost half of their time setting up experiments and taking readings and a quarter of their time talking about technical work.

These proportions vary from office to office, from job to job, as well as over time. Supervisors need to analyze each type of operation step by step as to what is being done and the flow of work. The current backlog may be compared with the current production rates in order to see how long it will take to complete all the work on hand. Daily, weekly, monthly, or seasonal variations can also be analyzed in order to enable better scheduling and smooth out workload fluctuations. There will be a gradually growing emphasis on personnel, meetings, guiding, showing, listening to others, reading about, or looking for, as well as applying new materials, tools, or other technological advances.

It is a highly recommended exercise even for every individual to do this for himself or herself. In many cases, the results are most revealing and can by themselves significantly help improve productivity. For the unit to be measured, this approach reveals on what most of the time is being spent and helps in setting productivity improvement priorities.

The activities and processes that take a great deal of time or resources quickly become apparent. Measurement should be directed toward the expected highest payoff areas, those areas where major problems seem to exist, which account for significant portions of the resources used, and where scarce resources are needed. To quote an example, in the health-care field in hospitals, apart from administration, the greatest opportunities for productivity improvement tend to be in those departments in which the major decisions are made by physicians, for example, operating room, emergency room, laboratory. These are the areas where productivity measurement is particularly needed. In contrast, most measurement to date has tended to relate to the accommodation/hotel aspects of hospitals.

Most productivity improvement actions can usually be taken at the organization's subunit level, such as in production, service provision, marketing, personnel, analytical functions, or administration. Overall productivity of the entire organization can be improved mainly by improving the productivity of any or all of these subunits. The aggregation of various subunit productivity levels or trends is usually not necessary. Aggregation always causes technical measurement problems and may not yield meaningful data at higher levels. Furthermore, at the organizationwide, corporate levels, where executives and senior managers have to make decisions, most of the major problems to deal with tend to be policy questions. The overall productivity improvement is usually the result of many lower-level productivity actions but needs to be stimulated and coordinated by top management.

Many tasks serve internal clients or users. For measurement purposes outputs need to be defined as identifiable, discrete output units irrespective of whether they serve internal or external clients. If the value of tasks serving internal clients needs to be determined, comparable tasks and outputs can often be found in the commercial private sector. Once the functions, outputs, and inputs have been identified, the relevant measures need to be chosen for output and input indicators.

Each of the chosen productivity measures—as has already been pointed out—need to be expressed in the form of a ratio of outputs divided by inputs. Productivity measures need to be in "real" terms, that is, both outputs and inputs must be measured in physical terms or proxy equivalents. Productivity ratios "in physical terms" include "constant" values. Examples of the relevant types of measures include the number of products of certain types produced, the number of products delivered, the number of certain types of services provided, the number of cheques issued, files searched, grant decisions made, people vaccinated—each of these per person-hour or per unit of capital equipment.

In the goods- or service-producing areas of the private sector, outputs can often be measured only in value terms, but if the productivity measurement is to cover a period of time, these must be adjusted to eliminate variations in the value of the currency in which they are measured. In the public sector and in many offices in the private sector most output measures tend to be physical measures. Inputs in these cases are mostly labor inputs, and these can be expressed in the form of labor time. In those cases of public-sector outputs where different products or services are for some reason measured in value terms, the same measurement processes apply as in the private sector, namely, expressing the values in real terms.

The appropriate procedures for the calculation of productivity measures will be discussed in detail in the following chapters. These will include the methodology relating to the measurement of outputs and inputs, that of interunit-level variations at one point in time, the preparation of productivity trend indexes, as well as the treatment of the effects of product-mix differences and inflationary influences.

It is important to point out that the meaning, applicability, and limitations of productivity indicators must be made clear. Whenever a productivity measure is used, it should not be called just "productivity" because of the existence of many kinds of productivity measures of both productivity levels (at one point in time) and productivity trends (over time). The specific measure used should always be described in very specific terms in order to avoid the possibility of misunderstanding and misinterpretation. For example, it should be as specific as "dozens of jeans produced per operator-hour," "number of patients treated per person-hour or person-day of a physician (or nurse)," or the "number of claim cheques issued per operator-day."

Although many of the required raw data are already available in existing company or agency accounts, they need to be adjusted for productivity measurement purposes, and the missing data that may still be needed should be collected. Qualitative bits of information that may not be presently used or even currently available can often be collected from client or patient feedback for future analysis. For example, many data are now available in caseworker records (for example, medical case openings, closings) but may not be utilized for productivity measurement and analysis at the present time.

Once the area for which productivity is to be measured has been defined, the appropriate productivity measure or measures should be chosen. The employees to be affected should be involved in the selection of the measures, particularly in the definition of the output to be measured. This is recommended because the productivity improvement action will be implemented by those employees and, in order to ensure their cooperation, they must understand and accept the measures.

# ANALYSIS OF THE PRESENT METHODS OF OPERATION

Improving the productive performance of an operation must start with a step-by-step diagnosis of the present operation: what is being done, why it is being done, how it contributes to the output of the organization, and how it is being done. Once we know where we are, we can set out to go where we would like to be and decide how to get there.

When diagnosing the present operation, first we need to define the desired output and set out clearly what we are doing. For example, are we running a machine or performing administrative services and, more specifically, what kind, for example, doing bookkeeping, maintaining accounts, collecting and recording accounts receivable, issuing cheques, searching for and selecting suppliers, processing purchase orders, verifying incoming supplies as to quantity, quality, and price, keeping inventories, or keeping records of maintenance schedules? Do we search for, gather, organize, and provide or distribute technical and/or management information? Do we administer human resources, for example, hire and fire people, keep performance records, or carry out appraisals? Do we manage communications, for example, receive, distribute, and send messages, process correspondence and reports, or perform a combination of a variety of such duties? How do any of these serve the organization's mission, goals, and objectives?

Measurements are needed for identifying and eliminating productivity and quality problems. Data can identify and help replace unjustified costly items, opportunities for performance feedback, and possible improvement actions. They help eliminate old accounts receivable and tie accountability to responsibility. Data also help identify and eliminate bottlenecks by better planning or scheduling of specific areas. Managers and supervisors can determine through measures what proportion of the problems or errors occurred due to poor planning or personnel problems, or the availability and quality of machines, tooling, or materials.

The key to acquiring the necessary information is keeping good records. Adequate and accurate record keeping is necessary for all output, as well as such inputs as direct and indirect labor, materials, and energy use, machinery, and physical distribution. Without the needed data, the managers and supervisors can not ensure that both the direct and indirect workers perform their assigned jobs in time and with satisfactory quality at the lowest achievable labor and other resource cost. Good records are also needed for customer relations. For example, in the hotel/motel industry guest accounts need to be regularly kept and reconciled with housekeeping records. Shortcomings must be corrected immediately, perhaps through additional training of personnel. Generally, it is also advisable that all employees know the cost of their activities in order to stimulate them toward higher productivity.

The productivity of the various operations needs to be measured, recorded, and analyzed systematically. In any productivity improvement program it is important to show early success in order to establish the credibility of the program. The managers should, therefore, not wait with implementing productivity improvements until complete productivity measurement systems are in place. Eventually, such systems will be necessary if they want to take optimal advantage of all productivity improvement opportunities, but significant success can be achieved early in the diagnostic phase of the productivity improvement program by implementing possible productivity improvements that have already been revealed.

Various regulations that had perhaps made sense some time ago but are no longer needed and cause unnecessary work or delays can often be spotted and eliminated by productivity analysis. Complex work can often be simplified and made not only less onerous and costly for the organization but also more pleasant and easier for the workers. These types of problems that hold back productivity improvement can often be identified and corrected right in the initial fact-finding, diagnostic phase by asking such questions as "Is this necessary, and if so, why?" or "Could it be eliminated, simplified, or done in some more effective or efficient way?"

In order to tell whether one is successful in meeting one's objectives, one must make comparisons with goals and objectives, past performance, or other standards or benchmarks, including those of competitors. Many organizations today learn from *benchmarking*. This simply means learning from comparing your unit to the performance and results of successful activities of other units/organizations for the purpose of productivity or quality improvement. The comparison is not necessarily made with competitors or even with other organizations in the same sector or industry, but from similar functions or activities of any successful organization in any industry. For example, successful accounting or physical distribution activities can be used as benchmarks or standards for any industry or company.

In view of such considerations, there is a general tendency to move toward the prescriptive use of indicators, setting objectives or targets against which performance can be measured. These are usually expressed in the form of standards against which performance can be judged. For instance, the use of standard times can help identify and eliminate superfluous activities in a more balanced manner, and improve scheduling and work flow.

After the initial diagnostic thrust, the identification of performance problems and further opportunities for productivity improvement must become an ongoing activity. The objectives need to be defined, production targets set, and the actual performance compared to the targets. For example, supervisors may compare the current number of people or machines on a certain operation with how many of each they actually need at a point in time. Unfortunately, the required labor and other inputs and their costs are too often not documented adequately. As a result of poor labor and other resource measurement and control, labor skills and necessary physical resources are often poorly utilized, leading to excessive input use and costs.

# QUESTIONS

Q: 3-1   Why is it a problem if you don't have appropriate productivity measures?

Q: 3-2   Is there a "best" or "right" productivity measure?

Q: 3-3   How can you know your productivity strengths and weaknesses?

Q: 3-4   How can you assess the relative importance of productivity and quality problems?

Q: 3-5   Give an example of how one can tell whether a company is weak in monitoring.

Q: 3-6   What must follow problem identification in order to achieve results?

Q: 3-7   If you are in office work, make a percentage breakdown of what you spend your personal time on during a typical workday. Use the list in A: 3.7.

# SUGGESTED ANSWERS

A: 3-1   The lack of appropriate productivity measures is a problem because without measures one does not know the facts, what one's strengths and weaknesses are, and how one performs. Without such knowledge, one can not take the right and effective corrective measures and improve.

A: 3-2    There is no "best" or "right" productivity concept or measure because productivity can be observed from many angles and therefore can be examined in many ways. For example, labor productivity, capital productivity, and other characteristics can be measured of the same operation, each of which is very important for various purposes.

A: 3-3    In order to determine your productivity strengths and weaknesses you need to measure your productivity and its factors and compare them to standards. The standards can be your objectives, your past performance, or the performance of your competitors.

A: 3-4    In order to assess the relative importance of productivity and quality problems, it is useful to list the problems in writing, rearrange them in order of priority, and take action to eliminate the most important ones in order of priority.

A: 3-5    Weakness in monitoring is indicated, for example, if the company can not provide accurate information on the time required for major phases of their operations or of the total person-hours necessary for the major phases of the operation.

A: 3-6    In order to achieve results, problem identification must be followed up by effective corrective action. This sounds obvious but is too often neglected.

A: 3-7    I spend approximately the following percentages (at least five percent) of my personal time during a typical workday:

- Reading mail                                    ____%

- Writing letters                                 ____%

- Making phone calls                              ____%

- Taking phone calls                              ____%

- Talking to peers on technical matters           ____%

- Attending meetings                              ____%

- Preparing for meetings                          ____%

- Reading literature                              ____%

- Gathering data or other information             ____%

- Entering records                                ____%

- Searching for and retrieving records        ____%
- Writing reports or memoranda                ____%
- Providing information on request            ____%
- Distributing information                    ____%
- Other official responsibilities (describe)  ____%
- Other, not directly productive activities   ____%

It is also very informative and useful to list the above items in order of importance (1, 2, 3, 4 . . .), number 1 being the most important.

# 4

# Developing
# Productivity Measures

## BASIC STATISTICAL CONCEPTS AND
## THE PRESENTATION OF DATA

Computer software packages can be used for many productivity and quality assurance operations, but virtually all workers today need to understand the data needs of their job and work unit, simple statistical concepts and methods, as well as basic statistical control processes. With such knowledge, they can feed the appropriate data and instructions into their computers and interpret the computers' output. Although most of us have learned basic mathematics, many readers may want to review the most important statistical concepts and methods.

Within the framework of elementary statistics, all employees need to understand, for instance, that the average of the numbers in question is the arithmetic mean. The *mean* is obtained by summing up all values and dividing their total by the number of items (that is, $3 + 4 + 7 + 9 + 15 = 38$; $38/5 = 7.6$. 7.6 is the "mean," or *average*.) All the items in question together are called the *population*.

It is often useful to obtain the *median* of a set of data. The median is the central element in the middle of the distribution. (In the above example, the median is the number 7, with an equal number of elements on either side of it in the distribution, that is, one-half of them are larger and one-half of them are smaller than the median.) The most common, the most frequently occurring value is called the *mode*. If, for example, a survey shows that a widget costs $1.75 in six stores, $1.98 in 20 stores, and $2.09 in nine stores, the most frequent value or mode is $1.98.

When products are produced or services provided, there is always some random variation around the standard characteristic from one item to the next. While, for example, the standard of certain pills manufactured is five gram, some are a bit heavier and others a little lighter. Some might be a little more acidic while others are less acidic. The variation of a characteristic around a central tendency, for example, around an average or a standard, is called the *distribution, dispersion,* or *scatter* of the characteristic.

A simple measure of dispersion is the *range,* which is simply the absolute difference between the lowest and highest values. The range is a significant tool of quality control because it is easy to understand, easy to calculate, and can be a very important bit of information when the variation around the mean is significant. The main weakness of the range measure is, however, that it is greatly affected by one extreme element and it is, therefore, not typically the preferred measure of dispersion.

The most important measure of dispersion is the *standard deviation* because when a large number of elements are measured, a predictable number of elements will be smaller and a predictable number of elements will be larger than the mean and because the proportion of the elements being at specific distances from the mean can be predicted with great accuracy. This distribution is called a *normal distribution.*

The absolute deviations from the mean, above and below the mean, add up by definition to zero (that is, the plus deviations and the minus deviations cancel each other out) and therefore we calculate the average squared deviation around the mean, which is always positive. The *average squared deviation* of the measurements from their mean is called the *variance.* For a total population, it is indicated by the symbol $\sigma^2$. The square root of the variance is the standard deviation.

The standard deviation is calculated by taking the difference of each item from the mean, squaring each difference, summing up the squared differences, dividing the total by the number of items, and taking the square root of the average of the squares. (There are various ways of performing this calculation, for example, from grouped data or ungrouped data, but computer programs are normally used for these calculations.) It is known that in a normal distribution 68.26 percent of all cases will be within one standard deviation.

For analytical purposes, the data can be presented in tables, textual form, or as graphics. It is helpful to display the data graphically, in *charts* or *graphs.* The simpler and often-used charts and graphs include the *frequency distributions,* in which numerical data are arranged according to magnitude, size, or other variable characteristic. The scale of the frequencies is shown on a vertical axis, while the class intervals are shown on the line of the horizontal axis. (The vertical line is called the *Y* axis, and the horizontal line the *X* axis.)

Frequency distributions can be presented in the form of a *histogram* or *bar chart*, in which the height of each rectangular bar shows the frequency of elements in each *class interval* represented by the bar.

*Scatter diagrams* show the distribution of individual items in a form where—similarly to the bar chart or frequency polygon—the scales are shown on perpendicular vertical and horizontal lines, respectively, and define the values of each individual point. For example, the vertical scale may show the weight of each machined part while the horizontal scale may show the time when the part was produced. This would show whether the weight of the products is getting larger or smaller over time. In another type of scatter diagram, the quality of each product may be indicated by how high it is placed on the vertical scale, while the pay scale of the worker who produced it may be shown by the item's location as measured on the horizontal scale. The relationship between the quality of products and the level of pay of the workers producing them would be clearly reflected by such a scatter diagram.

Statistical methods, such as the above, are very important in pinpointing trouble spots or evidence for proving that process control was maintained. Management and supervisors, members of quality circles, and even individual workers need to use such statistical methods when measuring the quality of incoming inputs and outgoing products. Statistical methods are used to control the actual production or service process, identifying the sources of trouble so that they can be corrected before many substandard items are produced.

Statistical methods are the only practical way of handling large volumes of numerical data and ensuring the high quality of production or services. These are carried out through the measurement of quality, data feedback, comparison with standards, and by helping to take corrective action when indicated. All information needs to be brought together, combined with the personal experience and judgment of managers and supervisors, and then it must be decided whether corrective action is both needed and justified in terms of expected benefits and costs with regard to both long-run and short-run considerations.

# MEASUREMENT OF QUALITY AND PROCESS CONTROL

Many of the measures of performance, productivity, profitability, and other analytical ratios by themselves reflect quality implications. Some examples of relevant measures are output per employee, value added per employee-hour, output per raw materials, production labor cost per hour, the relationship of output to fixed as well as current assets, machinery and equipment per employee, age of equipment, accuracy of equipment, capacity utilization, maintenance data, fixed costs and variable costs per product or service output unit.

Inventory ratios also can be useful indicators of the quality of inputs, such as the rework necessitated by rejects or other quality problems. The ratio of manufacturing or service overhead to value of output can point to quality shortcomings or, on the other hand, point to high-quality service. The various ratios of output to operating expenses or their elements such as selling, promotion, or administration also tend to be very useful quality indicators. For example, high overhead or selling expenses may—although not necessarily—indicate high-quality service, while low overhead or selling expenses may mean poor-quality products or service. It must be remembered, however, that expensive goods or services are not necessarily of high quality, and inexpensive products or services are not necessarily of low quality.

Once the quality control priorities have been selected, data are needed to ensure that a product or service conforms to the prescribed standard. As we have already seen, this conformance is the measure of quality. High or low quality means conformance or nonconformance to the specified standards (at the specified price). A simple, practical tool of statistical quality control (SQC) is the *Pareto analysis,* which leads to the identification of the primary causes of defects. Figure 4.1 shows a Pareto analysis.

In this example, the top two or three problem areas are the "vital few" that cause the overwhelming majority of the defects. Typically, solving the

| Defect type | Quantity all defects | Percent of all defects | Cumulative percent |
|---|---|---|---|
| Wood quality | 80 | 38.5 | 38.5 |
| Machine upkeep | 53 | 28.0 | 66.5 |
| Assembling | 45 | 20.0 | 86.5 |
| Upholstering | 12 | 6.5 | 93.0 |
| Turning | 8 | 5.5 | 98.5 |
| Training | 2 | 1.5 | 100.0 |
| **Total** | 200 | 100.0 | 100.0 |

**Figure 4.1**    Illustration of a Parato analysis.

vital few problems will greatly increase productivity and quality and, therefore, these problems should receive primary attention.

In the figure, 80 defects were found in "wood quality." This represented 38.5 percent of all defects. There were 53 defects in "machine upkeep," or 28 percent of all defects. Therefore, as the cumulative column shows, 66.5 percent of all defects were found in the top two activities. The Pareto analysis suggests the ranking of improvement opportunities. In this example, therefore, solving only the wood quality and machine upkeep problems will solve 66.5 percent of all the problems, that is, the majority of the defects.

There are various statistical methods that are helpful and commonly used in identifying *in control* and *out of control* process conditions. These can be illustrated on statistical process control charts from which the *common causes* of variability and the *special causes* of variation can be identified. The common causes are the random variations inherent in the process while the special causes are intermittent and unpredictable variations that need to be brought under control.

One type of the most commonly used process control charts is the so-called $\bar{X}$ chart, shown in Figure 4.2. The figure shows an $\bar{X}$ chart that consists of three horizontal lines, and of dots or asterisks that represent the individual samples. The chart is constructed within the framework of a vertical and a horizontal scale. The vertical scale usually shows the standard deviation from the population mean, in the same units as the data, that

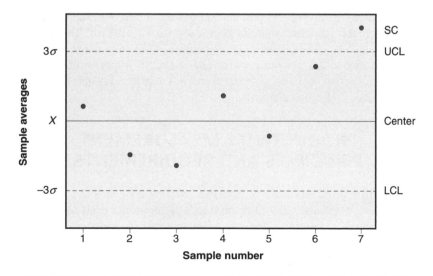

**Figure 4.2** Illustration of an $\bar{X}$ chart.

is, inches, grams, and so on. The center horizontal line is at the population mean. The upper and lower horizontal lines are the upper control limit (UCL) and lower control limit (LCL). These limits are typically set at three standard deviations from the population mean.

The averages of samples of a few, say five, 10, or 25, observations are represented by a dot or asterisk for each sample. The dots or asterisks are placed in relation to the mean, the upper control limit (UCL), and lower control limit (LCL). The population mean and the upper and lower control limits are calculated from the long-term process variance from historical observations when the process was "in control." In our example, dot no. 7 (SC) shows that the quality of this sample is outside the minimum standard (LCL–UCL) required. For example, the parts in sample no. 7 may be too heavy to meet the required standard. It is, therefore, a "special cause" of variance that needs to be investigated.

A significant benefit of statistical process control methods is that they require constant and systematic record keeping that tends to reveal cause-and-effect relationships and greatly helps in eliminating quality problems. Furthermore, since the record keeping involves the operator, it makes the operator aware of his or her own impact on the quality of production or service. It may also indicate to the worker (as well as to his superior) how better performance can be achieved in the future. It is important, however, to keep only measures that are needed for specific objectives. It is also useful to remember that raw data "keep better" over time. As the objective changes, the raw data can always be manipulated to serve current objectives.

As many employees as possible should be involved in the statistical control process; it is important to plan it together with them. The process flow needs to be charted and decisions made as to the areas that should be given priority in the statistical control process. Critical performance characteristics need to be given preference, and it is usually advisable to include historically problematic elements.

# PRODUCTIVITY MEASUREMENT PRINCIPLES AND REQUIREMENTS

In order to develop productivity measures, all data must be "observed" data rather than "standard" data. The productivity measures need to be compared with preset standards, forecasts, other comparable units, or compared over time with a base period. If appropriate, both efficiency and effectiveness measures, productivity (volume) and profitability (value) ratios, need to be developed. A variety of measures are needed to show the various aspects of the production or service. Usually several measures are used at the same time, in connection with each other, in order to show the different characteristics and causes of productivity levels and variations.

The theoretically desirable productivity measures should reflect volume relationships. In practice, however, value measures often need to be used as proxies for volume data. If value relationships are used over time, the changes in monetary value need to be excluded by use of suitable price indexes. One also must take into consideration that over time social valuations of different products also change. These mix physical output counts with values that are based on utility and market valuations.

Most of the productivity concepts use the concept of *value added.* This concept is often unfamiliar at the company level. "Value added" means the value added by the organization to the value of all goods and services purchased from outside the organization. For example, when raw materials are manufactured into products, the manufacturing process increases the value of the materials, "adds" value to the value of the raw materials. The difference in value is the value added. If both are expressed in *constant* dollars, they are called *real*, or *deflated,* value added (because the effect of inflation is eliminated). Value added can be measured in two ways, which give identical results: 1) the value of all goods and services purchased from the outside can be subtracted from the total revenue of the organization from the goods and services produced, or 2) by adding up the organization's internal costs, that is, labor cost (including management), capital cost, and profits.

For example, if all goods and services purchased from outside the company cost $40,000, and the total revenue from the goods and services sold is $100,000, the value added by the company was $60,000. According to the second way of calculating the value added by the company, the company's internal costs, that is, labor cost and capital cost together, amounted to $50,000, plus the $10,000 profit again makes the company's value added equal to $60,000.

Gross output can be related to a combination of various inputs or to a particular single input, such as labor or capital. We often talk about labor productivity throughout this book. Basically, it means output per employee or per employee-hour. As already emphasized, *labor productivity* does not mean and measure the specific contribution of labor to the output, but it measures how much output labor can produce with all the other resources, measured per labor time.

*Equipment productivity* can be measured by pounds, kilos, or litres of output per machine-hour, or by the number of units produced or processed per machine-hour. *Energy productivity* can be measured by units produced per cubic measure of gas or per litres of oil or per kilowatt-hours of electricity. *Multifactor productivity* measures usually combine labor and capital productivity. The so-called *total factor productivity* (TFP) relates gross output to a combination of all inputs, including labor, capital, materials and components purchased from others, energy, and purchased services, and so on, but TFP measures are rarely used at the company level.

In interfirm productivity and profitability comparison programs, usually about 30 to 40 different productivity and related ratios are calculated and provided to the users. (For the purposes of the analyst, as many as 100 quantitative and 100 qualitative indicators are often gathered and calculated, but these are used only for analytical purposes in order to find causal relationships and other factors that influence the main measures. However, these are usually retained by the analysts and only the main measures are presented to the nonstatistical user.)

Large companies often use many productivity indicators. For example, a very large manufacturing company is reported to classify the functions performed by its staff into about a dozen such "model function" categories as general services, personnel, maintenance, and so on, and more than a hundred activities, such as cost estimating, warehousing, secretarial service, and so on, each of which is associated with only one "model function." Productivity measures are calculated for these.

The choice and number of measures we want to use, therefore, will depend on the extent of our interest in and need for detailed productivity indicators. We should start with only a few measures. Successful companies have emphasized, however, that productivity measures based on financial and other quantitative information could well be supplemented by meaningful and well-organized subjective ratings by the users or customers of the products and services.

In order to be meaningful and meet their objectives, productivity and quality measures need to be:

- Relevant, reflecting the activities measured and fitting the organizational goals and objectives.

- Developed with some specific objective in mind. It is undesirable to collect more data and/or more often than necessary because this causes unnecessary cost. Too much data also wastes the time of the people who use it.

- Timely, in order to permit corrective action. They are of not much use if produced after the corrective action could have been taken or when the measures are no longer valid.

- Discrete, with identifiable results, tasks, and task elements.

- Accurate, reliable. Poor data may be misleading.

- From a sound data base and be consistent over time in order to permit tracking.

- Systematic, well organized, and fitting together.

- Based on outputs or activities that are measurable.

- Understandable, simple, even if several measures are needed. Productivity data need not be complicated. It must be kept in mind that most employees are not trained in statistical research techniques. Key output/input ratios, such as the number of pairs of shoes produced per operator-hour, per machine, or per unit of material, are all figures describing business facts and developments in a concise, understandable form, and are therefore useful to the decision maker.

- Practical, useful.

- Easy to administer.

- Cost-effective.

The frequency needed varies from case to case. Normally, more frequent measures are needed at the lower levels of an organization. For example, at the level of first-line supervisors, where the service is provided or the products are produced, daily measures may be the most suitable. Middle managers may need weekly measures while agency or company managers may need only monthly reports.

The output/input ratios can be developed in various details. For example, in storage and warehousing they can be calculated separately for purchasing, scheduling, receiving and put-away, storage, culling, replenishing, order selection, checking, order consolidation, packing, marking, loading, dispatching, shipping, overhead, and so on. The measures need to be detailed enough to permit the examination of relatively homogeneous units and identify the causes of variations or changes.

The productivity and quality measures include *quantitative* and *qualitative indicators*. The following are examples of quantitative productivity indicators: number of cheques, purchase orders, or invoices issued, stock requisitions filled or documents filed per clerk per day, drawings per engineer time, number of units of output per unit of energy used, number of units of output per units of materials input, or the (real) profits per employee or per payroll dollar.

Other useful, explanatory quantitative productivity indicators include the ratio of line personnel to staff personnel, direct workers to indirect workers, support personnel to total personnel, as well as secretarial support ratios, for example, the number of secretaries per the number of other types of staff. The quantitative productivity indicators also include such causal ratios as the number of errors per page of typewritten reports or per square unit of drawing and the engineering or drawing change rates per year.

Meaningful measures of work in process (WIP) include 1) WIP as percent of total materials, 2) value of WIP as percent of total production cost, and 3) WIP as percent of total output. Quantitative productivity measures in the administration of physical distribution include:

- Orders, units, lines, value of throughput per clerk-hour

- Stock location recorded/clerk time

- Cost of labor and equipment records kept/total cost

- Job tickets processed/labor hour

A retail establishment probably needs to keep and examine merchandise records of the volume and value of purchases (with dates), inventories, sales, mark-ups, and mark-downs for each type of merchandise. These data will enable the merchant to design and follow a plan showing how much of each stock should be carried at the end of each period, broken down by types, materials, size, price line. Such a record will enable the merchant to ensure that the store is never out of staple items.

Among qualitative productivity indicators client satisfaction with the services or goods provided is very important. These may result from unsolicited feedback from users of the organization's services or products, the organization's image in the eyes of clients or customers, and the number of repeat requests for the services or products of the unit. Examples of qualitative productivity indicators were given in Chapter 1.

# STEPS IN GATHERING ESSENTIAL DATA

Successful organizations agree that one of the key prerequisites of success is an extensive and purposeful use of data collection and analysis for the control of quality. By way of example, it is necessary to keep records of the use and condition of machinery and equipment in order to be able to carry out regular preventive maintenance. A record of equipment downtimes shows whether preventive maintenance is needed and minimizes unnecessary breakdowns.

The underlying data need to be collected systematically, regularly, and accurately. The principle is to measure in such detail that it is less costly to measure than not to measure. In most cases, it would be too costly or even unfeasible (for example in destructive testing) to measure all purchased or produced items. In such cases, *sampling* is used, that is, only a certain proportion of all items bought or produced are inspected. The meaningfulness of sampling is based on the above-mentioned finding that in a normal distribution, the sample will represent the total population with predictable accuracy.

In any sampling, there is a risk that the sample drawn from the total population is not representative of the total, with the result that a good lot may be rejected on the basis of the sample or, vice versa, a bad lot may be accepted. A sampling plan needs to be designed, therefore, to determine how large a sample should be taken and how much variation is allowed in the sample in order to meet the organization's quality objectives. When a sampling plan is decided for variables characteristics (which are measurable), for example, weight or chemical composition, the sampling decision is usually based on the normal distribution. If the mean of the sample and the range of variation within the samples are acceptable, the lot is accepted; otherwise it is rejected, indicating the necessity of further sampling or even the need for 100 percent inspection. In the case of a sampling plan for attributes characteristics (which are either "good" or "bad" and are, therefore, countable), the lot is accepted if the number of acceptable items meets the percentage prescribed in the sampling plan. Otherwise the lot is rejected.

There are two major sampling approaches to quality control, namely *acceptance sampling* (of incoming materials and parts and of outgoing products) and *process control*. The purpose of acceptance sampling is to estimate the proportion of defective items before or after the process is completed in order to ensure that on the basis of sampling, lots (or shipments) of a desirable quality level are accepted, and lots of an undesirable quality level are rejected. Such advance knowledge enables the producer to make the necessary corrections and allowances.

Acceptance sampling involves two kinds of risks. The producer or supplier risks drawing a sample that has a higher proportion of defective items than the total lot. In this case, a good lot is rejected, and producers hope to keep this risk low, usually below five percent. Consumers or other buyers risk, on the other hand, drawing a sample that has a smaller proportion of defective items than the total lot, resulting in the acceptance of a bad lot. Consumers try to keep this risk low.

The sampling plan needs to ensure that sample size will minimize the risk of both producers and consumers. The quality level of a good lot is called *acceptable quality level* (AQL), which is the maximum percentage of items in an inspected sample lot that can be defective and yet have the total shipment acceptable. The producer wants this level of quality accepted. The quality level of a bad lot is called *lot tolerance percent defective* (LTPD), which is the percent defective for an inspected sample lot for which the specific sampling plan gives a small probability of being accepted. The consumer wants to refuse a lot at this quality level.

The risk at the acceptable quality level (AQL) and at the lot tolerance percent defective (LTPD) level determines what the sample size and acceptance number or percentage of defectives need to be (between the AQL and LTPD points). The appropriate sampling plan, combining sample size and acceptance number of defectives, can be developed through trial and error or can be found in published standard plans.

The other major sampling approach to quality control, besides acceptance sampling, is *process control* of the actual production activities. Samples taken during the production process alert the operators, through, for example, a process control chart, when the process is out of control or moves in the direction of becoming out of control.

In order to have continual access to necessary information regarding managing a business and improving productivity, it is very helpful to use online information including all Internet sources, the public library, and governmental information sources, as well as to subscribe to at least one general business periodical and the leading industry trade media. For example, in advertising it is desirable to keep a copy of the ad on file and to maintain a record of the cost of the advertising and of the estimated value of its impact.

# QUESTIONS

Q: 4-1   Suggest major expenditure items for materials and service inputs.

Q: 4-2   What is meant by a "weighted" measure?

Q: 4-3   What does "value added" mean?

Q: 4-4   How can one identify quality problems and opportunities?

Q: 4-5   How can you obtain measures of poor quality?

Q: 4-6   How can you monitor operations when large volumes of output are produced?

# SUGGESTED ANSWERS

A: 4-1    Major expenditures for materials and service inputs include:

- Materials inputs (for example, raw materials, goods purchased for resale, operating or other supplies, energy)

- Inputs of purchased services (for example, research and development, professional fees, work contracted out, communication, storage and transportation, travel and hospitality, advertising and sales promotion, repairs and maintenance, leasing and renting, royalty payments)

- Employment (salaries, wages, and supplementary labor income)

- Depreciation

A: 4-2    We need to use a "weighted" measure when we want to add up various outputs or inputs such as chairs and tables. Each of these must be expressed in common terms by using "weights." The weights can be unit labor requirements (ULR), that is, how long it takes for a worker to make each, or relative values such as prices. The "weighted" outputs or inputs then can be added together.

A: 4-3    "Value added" means the difference between the value of goods and services the organization bought from others and the value of goods and services sold by the organization, that is, the value "added" by the organization.

A: 4-4    Quality problems and opportunities can be identified by historical data on:

- Customer feedback and complaints

- Returning customers

- Surveys

- Checklists

- Certain high input costs

- Too large WIP (work in process)

- Frequent breakdowns

- Too long turnaround time

A: 4-5   Measures of poor quality can be developed by regularly keeping track of:

- Rework, reprogramming, retesting, or redelivery, and so on

- Customer complaints

- Returns

- Warranty costs, and so on

- Errors and defects

A: 4-6   When large volumes of outputs need to be monitored, work-saving methods of statistical sampling or control by exception can be used. Sampling only examines a certain percentage of products produced or services provided and draws overall conclusions as to the total output. Monitoring by exception demands a report only when the process deviates from the plan.

# 5

# Measuring Output and Input

## CRITERIA AND METHODS OF CHOOSING OUTPUT INDICATORS

The purpose of productivity measurement is to find out how we could produce the wanted goods or provide the wanted services from the minimal human effort and physical resources. The measurement of output and input is, therefore, an essential element of productivity analysis.

First, let us deal with output. There are many ways of measuring output, at various points in the economic activities, with different methods, as well as using various yardsticks. The key to successful productivity measurement is the appropriate selection of output indicators. The appropriate selection, in turn, depends on the very clear definition of what is being done. The objective of the selection is to ensure that the output measures should be meaningful reflections of the production of the establishment and that the technical characteristics of the output measures should be suitable for reflecting organizational productivity changes and variations. Therefore, the most important criteria of the useful output indicators for organizational productivity measures are highlighted in the following:

1. The output measure should cover a significant part of the establishment's production or service. The measure should be meaningful from the viewpoint of the establishment and provide a reliable indication of the relevant part of the output. It should be kept in mind that people try to meet the requirements of the measure used to evaluate their performance. The output measure to be chosen should, therefore, focus on the main objectives of the establishment.

2. The output measures should reflect the final product or service of the organizational unit. It should be final from the point of view of the unit being measured. The product produced or service provided should be consumed or used outside that establishment. This is important because many establishments produce outputs that are intermediate inputs of the larger organization to which they belong. If, for instance, the productivity of the entire organization is being measured, the output should only show the final product or service of the entire organization. All resources used should be related to that output. When, however, the productivity of, say, the personnel branch of an organization is measured, the output to be measured is the service that is considered final for that branch. Duplication of output reporting should also be avoided in organizations composed of head offices and regional units. If an output is the joint product of the head office and regional offices, it should be counted only once.

3. The output measures should reflect the workload of the organizational unit to be measured. As mentioned earlier, work done by contractors is not the output of the organization. In such cases only the work actually done by the organization's workforce, for example, the management and monitoring of the contract, the investigation, determination, and administration of the contract, inspections, follow-up visits, and similar contributions, are the contracting organization's output. The work done by the contractor is not the output of the organization giving out the contract. It is not the output produced from the input of the contracting-out organization.

4. The inputs and outputs must match in all important aspects, that is, coverage, time period, and so on. In order to ensure that inputs and outputs match:

    a. Opening inventories that were not produced by the current period's inputs should be excluded from current outputs. On the other hand, closing inventories made from current inputs need to be included in current output.

    b. Labor input hours must be hours "worked" or hours "at work," in real terms, for example, overtime hours need to be counted as straight time.

    c. Capital input needs to be recalculated in "constant" values and depreciated over the true *economic* life of the asset, that is, as long as it is being utilized. The *economic depreciation* over the current period—that is called *capital flow*—is the capital input for the period.

Various outputs or inputs can be totaled up by weighting each output (or input, respectively) by some common element. For example, when adding telephone calls answered and letters typed by a secretary, each phone call may take two minutes, but each letter may take 12 minutes. In this case, the number of phone calls should be multiplied by two and each letter by 12 so that they could be added and divided by the input, say, operator hours.

5.  The output measures should be independent from the inputs. For instance, output measures should not be based on the labor input required for their production because it would mean dividing something by itself. This rule is not inconsistent with the weighting of various outputs by their respective base year unit labor requirements. As we have already seen, such weighting is necessary, for example, when two types of products or services are produced by the same establishment and a unit of one of the products requires, say, twice as much time to produce as one unit of the other. Before they can be added together, each product must be multiplied, that is, weighted, by its respective base year unit labor requirement.

    Macroeconomic-type outputs in the public sector are often measured by inputs for national accounting purposes by assuming that the value of output is equal to the cost of the labor producing it. While these measures are used by convention for national accounting purposes, they are not suitable for productivity measurement purposes—even in the public sector.

6.  The output measures should reflect clearly identifiable and quantifiable production or service output units. The number of chairs produced, number of cheques issued, number of training classes taught, or the number of applications investigated are suitable examples of output measures.

7.  The units measured by each output measure should be relatively homogeneous as to their labor requirements. Assuming, for instance, that individual selection interviews and appraisals take about the same officer time in personnel branches, both can be included in the same output count. If, however, their relative unit

labor requirements differ, such as those of tables and of chairs, a combined measure would distort the productivity measures.

8. The outputs to be measured should preferably be repetitive. Outputs are easiest to measure if they are produced on a regular basis. They can, however, also meet this criterion if they are performed on a repetitive basis although on request only or at irregular intervals. For example, market information disseminated on request or health inspections in response to need meet this requirement. Large, unique projects can often be broken down into separate phases or tasks that in themselves can meet the requirement of repetitiveness.

9. Changes in quality should be reflected by output measures if they involve the use of more (or less) resource inputs. If, for instance, better service is provided than earlier through a more intensive analysis, the numerical count of output may show an apparent decline in efficiency when in reality a smaller number but better-quality services are provided. In such cases the comparison should be based on how much labor would have been taken in the base year to produce the improved service unit, and weighted accordingly. The main objective is to avoid adding apples to oranges in the output measures. In such cases, separate output measures should be designed for each output type. They can only be added up if weighted.

10. Output as well as input data should be measured on a consistent basis throughout the period covered. This is particularly important to observe when reorganizations take place. All data should be recalculated to represent a consistent series through the base year. For example, estimates should be made for a new office as to how much its output and input would have been in the base year if it had already been in existence. These data could then be used to calculate the productivity index after the reorganization.

11. The data should be gathered in sufficient detail to enable meeting the above criteria. For this purpose, records should be kept in appropriate detail, for example, separate data for product or service lines, customers, area, salespersons.

12. The less sophisticated the measures are, the more useful they generally are. The need is for something workable and understandable.

# IDENTIFYING PROBLEMS AND OPPORTUNITIES OF OUTPUT QUALITY

As a prerequisite to any consideration of the quality of the outputs of products and services, factual information must be obtained on the current and desired quality levels. It is necessary to perform several distinct steps:

1. Identification of the factors that need measurement and investigation. Factors that are not under the control of the organization or group measured should be excluded from quality measures.

2. Careful, systematic, and cost-effective collection of data.

3. Systematic assembling and classification of the data into meaningful categories, distributions, or series, for example, by magnitude or time of occurrence.

4. Analysis of the data in comparison with specific benchmarks such as standards, targets, or goals.

5. Finding out why quality variations occurred.

6. Determining the causes of variation.

7. Identifying how much of the total difference from the standard is due to what reasons.

8. Investigation to ascertain whether corrective action is necessary, and making corrections where needed.

9. Feedback to all concerned in order to ensure that repetition of the defects is avoided and that future performance is of high and constantly improving quality.

Specific aspects of quality are indicated (and should be weighted) by the various quality characteristics, such as comfort, durability, and so on.

One of the most important and useful sources for identifying quality problems is feedback from customers. Once the valid quality demands of customers are determined it is necessary to find out the reasons for possible dissatisfaction. Constant interaction must be maintained with them. There are many techniques for seeking and obtaining customer feedback concerning product or service quality. For example, a producer can carry out surveys of customers about their likes or dislikes of the product or service, gather information through warranty registration cards or self-addressed and stamped reply cards, follow up unsolicited comments received from customers, and gather information from purchasers or users of competitive products or services as to why they chose to buy that product or service.

Producers should talk to customers, listen to them and record, as well as promptly follow up on, their valid observations or suggestions. The comment cards or letters as well as the oral questions can ask the customers such questions as whether and to what degree they were satisfied with the product or service received, in what condition it was received, whether the service or delivery personnel were courteous, and whether they had any suggestions on how the service could be improved to better satisfy their needs.

A very important source of customer feedback on the quality of a product or service provided consists of customer complaints. These complaints reveal what the customer feels about the product or service. Customer complaints must be followed up carefully and promptly. In order to satisfy the complainers as well as all other customers, it must be ensured that the customer feedback reaches those who are responsible for correcting the defect or error.

Since there are many possible aspects of poor quality and, therefore, opportunities to make quality improvements, it is necessary to identify the causes of quality problems. The clarification of the major outputs and services itself often points to major potential problem areas. If historical evidence is available, it can indicate typical common defects. These can be included in a checklist for each process step. Self-checking techniques can be built into the production process. These by themselves would result in fewer defects. Charting the production process in detail also tends to reveal opportunities for errors or defects. Such charting can take the form of developing and maintaining flowcharts and process analyses. In general, statistical methods are very important in revealing where the fault lies.

The most important parameters for quality comparison are product specifications, performance, appearance, and safety. If comparisons with industry averages, competitors, or other similar operations are available, relatively high material costs can point, for instance, to a respectively high proportion of rejects, other forms of material waste, or poor purchasing methods. Relatively high labor input may indicate much waste of labor time through poor planning and scheduling, or poorly made output that makes reworking necessary. High operating expenses may reflect poor-quality production, unnecessary red tape, redundancy, or, for example, sloppy invoicing. An especially large amount of work in process may indicate poor planning or scheduling or a high inventory that is necessary due to too many reworks of faulty products. Low utilization of fixed assets may point to low-quality maintenance that results in frequent breakdowns.

If comparative information is not available, the unusual or relatively too-rapid growth of certain input categories may identify productivity or quality problem areas. Such growth data are very useful for indicating areas where quality is improving or deteriorating in comparison with other production elements. It has been observed that a few problem areas tend to account for the large majority of defects. This observation of the "vital few and trivial many" points to the need for concentrated attention on those areas that cause most of the productivity and quality problems.

The distance between an airplane's places of departure and arrival can be counted in terms of miles or kilometres, but the quality of the service provided by the airline is less obviously measurable. The answer lies in the dual nature of quality, in the measurement of both the physical quality characteristics of the product or service and the subjective characteristics of each.

The illustration of objective and subjective quality attributes of products or services suggests that quality needs to be measured by a group of measures, some countable, others quantifiable only in the eyes of the customer.

The above considerations raise the question of how to measure quality. While the number of pairs of shoes can be counted, their quality is more complex. In order to reflect the various aspects of quality, a variety of quality measures are needed. These should be described in the form of "valid," that is, justified standards and specifications desired by the customers. The standards and specifications must also take into consideration the resources used in the production of the goods or the provision of the services, as well as the price of the products or services. The "family" of quality measures can be analyzed and evaluated individually—with proper regard to the other related measures—or built into a weighted overall quality measure. The relevant indicators can be weighted according to the priorities of the customers. For example, measures derived from a user utilization survey and such indicators as the proportion of products or services delivered on time and budget may be given greater weights than such efficiency measures as the number of defects or errors per output of units produced or the quantity of products or services produced per hour. Companies aiming for high quality performance need to have this type of information available.

The first question is how many and what data should be collected for quality measurement and improvement purposes. Successful large companies collect literally hundreds of measures that enable their employees to monitor their performance. Smaller organizations don't need to gather so much data. Nevertheless it is essential for all organizations to keep sufficient data to enable them to know the quality of their performance and the sources of trouble in order to take corrective actions when indicated. The need for specific data varies from organization to organization, but there are certain guidelines that can help organizations choose and collect the most vital types of numerical and statistical information. Some of the most useful measures are briefly reviewed following.

Direct quality control data typically include inspection data of the quality of incoming materials and parts, the number of errors or delays in all external procedures (for example, time of order receipt, processing, delivery) and internal operations (for example, accuracy in order processing). In production, useful measures include output per employee, output per hour of work, the volume and value of scrap and returns, rework of defective units, retesting, readministration, warranty costs, customer satisfaction, market share, as well as employee morale, which could be measured by turnover, absenteeism, and employee surveys.

Measures of customer valuations of quality include the length of wait for the order to be filled, on-time delivery of products or services, the number of complaints, the time taken to satisfactorily resolve problems, reliability of the product or service to the consumer. Customer ratings may be classified by the various producing departments, for instance, design and engineering, production management, sales management, and so on.

The volume as well as cost of the products produced or services provided need to be measured and recorded because, among many other reasons, a number of quality indicators will be related to them. In order to be useful, it is necessary to define clearly and accurately what quality characteristics apply to each product or service. For outputs of products and services, therefore, it is not only the volume and value data that are needed but also their specifications, properties, and other characteristics such as performance and appeal to the customer or other user.

Quality defects can be introduced into the system by many causes, such as lack of specifications, faulty design, poor material, faulty procedures, inadequate timeliness, lack of adequate measurement or quality control, antiquated or unsuitable equipment, poor skills, job instructions or training, inexperienced or untrained employees, lack of management concern, poor administration, billing, or delivery.

Once the problem is identified, one must find and eliminate the cause(s) of the defect or error. If historical evidence is available, it can indicate typical common defects. Checklists of common errors or high-risk procedures can be developed by well-trained and experienced staff and made available to junior workers to use as a sort of technological or experience transfer to help prevent or find and eliminate the typical causes of quality problems.

# WHAT DATA ARE NEEDED FOR
# MEASURING OUTPUT IN SERVICES

It is often assumed—incorrectly—that productivity in the service sector can not be measured because "service outputs are hard to quantify." It is also thought that outputs in the public sector, which mostly consist of services, would be even harder to measure because public-sector services do not usually have price tags attached to them that would help in assessing the value of the output. In fact, most service-sector outputs can be measured at the level of the final output of organizational units. The key to finding suitable output measures is to define very clearly what is produced, exactly what is being done. These measures often include not only quantitative but also qualitative indicators that reflect, for instance, the speed of service, timeliness, accuracy, errors, backlog of work, or customer satisfaction.

In view of the vital importance of improving productivity in services much more rapidly—which has only been realized in recent years—conscious productivity improvement programs need to be undertaken. It is clear from past experience that self-generated productivity improvement is very slow but that conscious and systematic productivity improvement programs can bring rapid and substantial results.

An example is that of hospitals where quality variations at one point in time and over time can be reflected by the intensity and duration of care. The first would indicate such aspects as what was done to the patient and how often. Service duration is normally measured by the length of hospital stay. This may be adjusted by diagnostic group. The two measures, service intensity and service duration, tend to be inversely correlated. If the productivity of services is compared between hospitals, it is essential first to look at the differences in the demands placed on the various hospitals. The demand for services tends to vary significantly from region to region. In banks, the length of waiting in lines, the prompt and accurate provision of statements, and the results of customer surveys are typical measures of the quality of service output.

Perhaps one of the most difficult areas of measuring output for productivity purposes is research and development (R&D). In this area, the reaching of preset objectives, accepted standards, or peer recognition may be some of the best overall output indicators. Also, in practice, much of the R&D work consists of a variety of clearly measurable elements, for example:

- Designs prepared

- Tests carried out

- Experiments conducted

- New standards developed

- Instruments calibrated

- Percent of output identified as erroneous or redundant

- Percent of output questioned by customers/clients

- Proportion of output that meets standards

- Timeliness of project completion

- Peer evaluation

- Accuracy of predicting needs for space or facilities

In the case of projects that extend through long periods of time, distinction may be made between different steps, phases, activities, or individual tasks that may be measurable in themselves.

Another area the output of which is often assumed to be hard to measure is that of professionals, knowledge workers, and skilled workers in offices, a group that partly overlaps with those in research and development. In fact, measures for these outputs can usually be found. The most essential requirement in the development of output measures for professionals, knowledge workers, indirect workers, and skilled workers in offices is that it must be clearly defined what the indicators are trying to measure.

Examples of possible measures of office workers are the identification of discrete and identifiable tasks or task elements (their results, if possible) including timeliness. These can be compared with similar work done elsewhere, keeping the measures simple and meaningful. Fixed versus variable work should be identified. It should be measured how long it takes to do each task and how often. The sensitivity of the measures with regard to the outcome should be analyzed. The people whose group is measured should be involved.

A particularly urgent area for productivity improvement—and therefore its measurement—is that of the distributive trades because distribution and the various forms of transportation account for a very large part of retail prices to consumers. Productivity measurement in the insurance industry is also more and more important.

The best way to find appropriate measures for output for productivity measurement in this sector is to ask oneself, "What am I trying to do?" "Is this or that part of what my operation is doing as well as that of my competitors?" and "Is it done better or worse than last year?"

# THE MEASUREMENT
# OF INPUTS

The most common inputs used in productivity measures are labor input, capital input, or a combination of both. For this purpose it is necessary to determine how much labor and/or capital is required to get a job done, as well as how long and how often has it to be done. It is useful most of the time to distinguish between fixed and variable types of inputs.

The measurement of labor input is perhaps the easiest. The measurement of other inputs, such as capital, energy, and others, is much more complex. Furthermore, data on the inputs of capital and other inputs are much scarcer. Labor input is measured in physical terms, namely, labor time. It is usually expressed in person-hours, person-days, or person-years. For productivity purposes, the ideal labor input measure would be "hours worked" on the output. This is, however, rarely available except from special industrial engineering–type stopwatch studies or special tabulations. The measure of labor preferably used is, therefore, "hours at work," which includes the hours worked but also other components of "attendance" time, such as the time from entering the place of work until the actual start of the work, waiting time, meetings, training time, as well as time that may be spent at the workplace on irrelevant activities.

An alternate labor input measure is "hours paid." This is often used when "hours at work" data are not available. This measure is less accurate in reflecting the use of labor in producing the output because it includes time on sick leave, paid vacation, and holidays. However, as long as the proportion of time spent on these nonworking items remains stable over time, and the comparisons are made over time, the distortion caused by the use of "hours paid" instead of "hours at work" is not significant.

It is worth noting that the number of "hours at work" is usually available at the time when payrolls are prepared but these data are usually not kept on record by accountants. Their recording for future productivity use—perhaps by pencil on the margin of the ledger—is relatively simple and inexpensive, and should be encouraged. Some of the special problems of labor input measurement include the treatment of part-time work, the hours of salaried rather than hourly employees, related work performed by central agencies for the unit measured, and the qualification or composition of labor input.

Part-time labor input is usually converted into full-time labor equivalents so that they can be added to the full-time employees. For example, two half-time workers are counted as one full-time worker. Overtime work is counted as straight-time hours even if paid at twice or more the straight-time wage rate. Salaried employees are assumed to work a normal work week, for example, 37, 40, or other hours, as the case may be. The time of central agencies, for instance accounting or personnel, may be allocated to the service-producing units on the basis of, say, the proportion of the relative use of the central service. In other cases such allocation is not done in order to keep the direct input–output relationship clear. This has to be decided case by case.

It is important to note that all labor input is counted, including professional, managerial, as well as clerical and other supporting labor. The labor input is normally taken as homogeneous, mainly because the use of education, experience, salary, wage rate, or other characteristics is not clearly or consistently reflected in the volume or quality of outputs. For example, the salary of a worker is not necessarily—or even normally—proportionate to his or her productivity. The bias caused by this problem can be minimized by endeavoring to choose operating units at a level of detail where the labor input is relatively homogeneous.

The inclusion of capital input in productivity measures is significant, particularly in capital-intensive operations, say, manufacturing, computer work, some laboratory or other testing functions. Here, the "flow" of fixed capital should be measured rather than the "stock" of fixed capital. The "flow" of capital is the portion of capital deemed to be "used up" in the production of the particular output. If some equipment is, for instance, used for 10 years, the "flow" of capital in one year is around one-tenth of the cost of the full cost of that capital asset. The latter, the full cost of the capital asset, is called the "stock" of the capital.

If it is decided to include fixed capital in the productivity input, the various capital stock items should be expressed in constant values (by "deflation," that is, eliminating the changes in currency values over time), then divided by the number of years of expected economic life of the capital asset. This yearly capital flow can then be divided by the average hourly wage of the workers who produced that capital asset, which gives the capital flow in terms of "equivalent hours." The equivalent hours can then be added to the labor hours for the combined multifactor productivity ratio. When the capital asset is imported, the "equivalent" hours can be calculated by using the hourly wage rate of the workers who produced the asset in their country. The import cost of the asset can be divided by the wage rate in order to yield the equivalent hours.

"Current assets," that is, inventories and accounts receivable, are not normally significant enough in offices to be included in productivity measures. In manufacturing and similar operations, however, they may affect business performance significantly. If they represent a significant part of costs, they should be minimized.

# QUESTIONS

Q: 5-1   What is the difference between productivity measurement and conventional accounting?

Q: 5-2   List a manufacturing company's main fixed-asset inputs.

Q: 5-3   What are a restaurant's main inputs?

Q: 5-4   Suggest some measures of the quality of service in physical distribution.

Q: 5-5   How can you total up various quality measures?

Q: 5-6   List some of the causes of quality problems.

Q: 5-7   Give examples of quality in personnel work.

Q: 5-8   What does quality mean in services?

# SUGGESTED ANSWERS

A: 5-1   The difference between productivity measurement and conventional accounting is that productivity measures express the relationship between the "real" resources used and the "real" output of products and services. Taxation laws, conventional accounting practices, and the ownership of the assets used do not affect productivity relationships. However, with proper adjustment, accounting data can be used for developing productivity measures.

A: 5-2   A manufacturing company's major fixed input cost elements include:

- Land and buildings

- Machinery and equipment

- Road vehicles

- Electronic equipment

- Furniture and fixtures

A: 5-3   A restaurant's main inputs include the cost of sales (food and beverages) and controllable costs, such as:

- Administration and general

- Labor (payroll and fringe benefits)

- Direct operating expenses

- Music and entertainment

- Repair and maintenance

- Utilities

- Advertising and promotion

- Depreciation

A: 5-4   Examples of quality of service in physical distribution include:

- Response time

- Proportion of on-time deliveries

- Job rejects per total jobs

- Scrap quantity per total inventory

- Setup time per run time

- Clerical time per order or unit

- Number of inquiries left unanswered per day

- Number of days taken to resolve valid customer complaints

A: 5-5   In order to total up various quality measures, one has to weight them by the relative importance of each, by the relative unit labor requirement (ULR) of each, or by the relative unit cost of each.

A: 5-6   Quality problems can be caused by such matters as inadequate instructions, procedures, or qualifications, the lack of management concern, or lack of control.

A: 5-7   Examples of quality in personnel work include confidentiality, fairness, responsiveness, timeliness, accuracy, empathy, and helpfulness.

A: 5.8   In services, quality means such characteristics as timeliness, quick response, usefulness, accuracy, or meeting objectives.

# 6

# Comparing Productivity Indexes of Two Organizations

## METHODS OF CONSTRUCTION

It is often necessary to construct productivity indexes in order to reveal how productivity of a company performed in comparison with another, or whether and to what extent productivity has improved or deteriorated. Basic methods of index construction are, therefore, explained in the following.

### Calculation of Productivity Differences in Percentages, Based on Absolute Data, When Outputs Are Homogeneous

It is rare that an organization produces only one kind of relatively homogeneous, that is, more or less identical, product or service. If this is the case, however, productivity measurement is rather simple. For instance, if organization A produces five letters per word-processor-hour and organization B produces four letters per word-processor-hour, the productivity of organization A is 25 percent higher than that of organization B. Example 6.1 shows the required calculation. It is illustrated by Figure 6.1.

*Explanation of Example 6.1.* Organization A produces 15 letters by three word processors in one hour. Its productivity, therefore is $15/3 = 5$ letters per word-processor-hour.

Organization B produces 28 letters by seven word processors in one hour. Its productivity is $28/7 = 4$ letters per word-processor-hour.

As it was noted earlier, the productivity of an individual or organization is always compared to some "standard" or "base" to see how much better or worse the compared unit is performing than the standard or base, that is, the

## EXAMPLE 6.1

### Calculation of Productivity Differences in Percentages, Based on Absolute Data, When Outputs Are Homogeneous

|  | Output of letters processed (a) | Input of word processor hours (b) | Productivity (output per word processor hour) (c) | Productivity index % (d) |
|---|---|---|---|---|
| Organization A | 15 | 3 | 5 | 125 |
| Organization B | 28 | 7 | 4 | 100 |

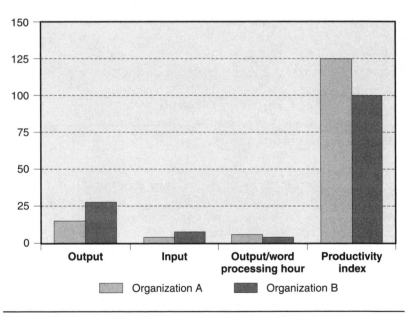

**Figure 6.1**    Productivity differences in percent from absolute data, homogeneous output.

same unit in the "base period." The comparison is usually done in percentage terms. The units to be compared are converted into index form, for which the standard or base is taken as 100 percent.

In Example 6.1, the productivity of organization B (4) is taken as the standard with which the productivity of organization A (5) is compared. The index of organization B (4) is the standard 100 percent. The index of organization A is calculated by dividing its productivity (5) by the productivity of the standard (4) and multiplying the result by 100, yielding (5/4) × 100 = 125. To obtain the percentage difference between the productivity of organization A and organization B, the standard is deducted from the index of organization A, 125 − 100 = 25.

## Construction of Productivity Indexes from Raw Productivity Data, and Calculation of Productivity Changes over Time When Outputs Are Homogeneous

When an organization produces the same homogeneous output over time and we are interested in finding out how its productivity changes, the calculation is similar. Using the data of organization A, the required calculations are shown in Example 6.2 and illustrated by Figure 6.2.

*Explanation of Example 6.2.* In year 1, the organization produces 15 letters per hour with three word processors, yielding a productivity of five

---

### EXAMPLE 6.2

#### Construction of Productivity Indexes from Raw Productivity Data, and Calculation of Productivity Changes over Time When Outputs Are Homogeneous

| Year | Output of letters processed (no.) (a) | Input of word processor hours (b) | Productivity (output per word processor hour) (c) = a/b | Productivity index (Base year = 100) (d) | Productivity change from previous year % (e) |
|---|---|---|---|---|---|
| 1 | 15 | 3 | 5 | 100 | — |
| 2 | 22 | 4 | 5.5 | 110 | 10 |
| 3 | 30 | 5 | 6 | 120 | 9 |

Note: The data for productivity change have been rounded.

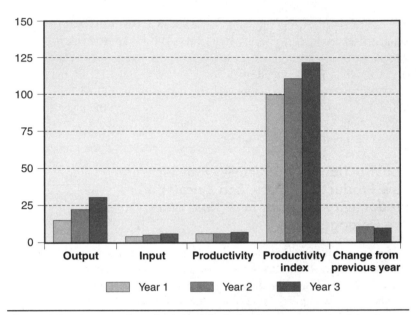

**Figure 6.2** Construction of productivity index from raw productivity data, homogeneous outputs.

letters per word-processor-hour. This "base year" productivity is now the "standard" or "base year" index of 100.

In year 2, 22 letters are processed per hour by four word processors, yielding a productivity of 5.5 letters per word-processor-hour. The corresponding index is obtained by dividing the productivity of year 2 (5.5) by the productivity of year 1 (5) and multiplying by 100, yielding a productivity index of 110. This is 10 percent higher than in the base year.

In year 3, 30 letters are produced per hour by five word processors, yielding a productivity of six letters per word-processor-hour. Dividing this figure (6) by the productivity of the base year (5) and multiplying by 100, the index of 120 is obtained, meaning that the productivity achieved in year 3 is 20 percent higher than that in the base year.

If we want to find out how much the productivity has increased in year 3 from the preceding year (year 2) rather than from the base year (year 1), the index of year 3 (120) is divided by the index of the preceding year (110), yielding 109. As this comparison has been made with the preceding year (year 2), it is year 2 that has to be taken as the standard 100 percent, and if this (100) is subtracted from the 109, it is seen that the percentage increase from the preceding year is nine percent.

## Construction of Productivity Indexes from Output and Input Indexes When Outputs Are Homogeneous

The same productivity indexes shown in Example 6.2, column (d), can also be calculated directly from the raw output and input data without first calculating the actual productivity figures for each year. In this case both the output and input data are converted into indexes by dividing each by its own base year figure and multiplying it by 100. Then the productivity index for each year is calculated by dividing each year's output index by the same year's input index and multiplying the result by 100. This procedure is shown in Example 6.3, and illustrated by Figure 6.3.

***Explanation of Example 6.3.*** The output index of year 2 is obtained by dividing the raw output figure of year 2 (22) by the base year output figure (15) and multiplying the result by 100. This yields an output index of 146.7 for year 2. For year 3, the raw output figure of year 3 (30) is divided by the base year output figure (15) and the result multiplied by 100. This yields an output index of 200 for year 3.

Next, input indexes (column d) are calculated by the same method from the raw data (column c). Finally the output indexes (column b) are divided by the corresponding input indexes (column d) and the results are multiplied by 100 to yield each year's productivity index (column e).

---

### EXAMPLE 6.3

#### Construction of Productivity Indexes from Output and Input Indexes When Outputs Are Homogeneous

| Year | Output of letters processed (no.) (a) | Output converted into index (Base year = 100) (b) | Input of word processor hours (no.) (c) | Input converted into index (Base year = 100) (d) | Productivity index (Base year = 100) (e) = (b/d) × 100 |
|------|------|------|------|------|------|
| 1 | 15 | 100 | 3 | 100 | 100 |
| 2 | 22 | 146.7 | 4 | 133.3 | 110 |
| 3 | 30 | 200 | 5 | 166.7 | 120 |

Note: The data for productivity change have been rounded.

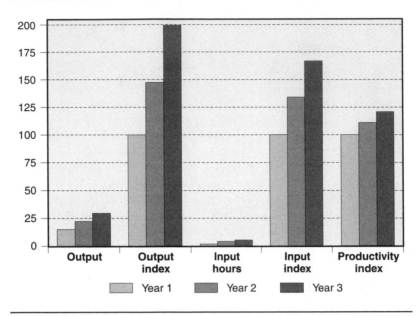

**Figure 6.3** Productivity index constructed from indexes, homogeneous outputs.

## Calculating the Productivity of Two Organizations with Heterogeneous Products Using Unit Labor Requirement (ULR) Weights

The above examples showed the construction of productivity indexes for organizations that produce only one kind of relatively homogeneous product or service. In reality, most organizations produce various outputs, in various combinations. By way of example, one organization processes many letters and has few phone calls, while another organization processes about the same number of letters but handles many phone calls.

Example 6.4 shows the basic data used for the calculation of *unit labor requirements* (ULRs) in Example 6.5, and information from both Example 6.4 and Example 6.5 is needed for the calculation of weighted output indexes in Example 6.6. The calculation of input indexes is shown in Example 6.7.

As it usually takes much longer to process a letter than to refer a phone call, we must adjust or "weight" the number of letters and phone calls appropriately if we want to compare the productivity of the two organizations. The "weighting" should be done by an appropriate and measurable relative characteristic of the compared outputs, for instance by their respective unit

## EXAMPLE 6.4

### Basic Data for Calculating Weighted Productivity Indexes over Time

|  | Output | Person-years input |
|---|---|---|
| **Year 1** | | |
| Output A | 10 | 21 |
| Output B | 15 | 6 |
| Total | | 27 |
| **Year 2** | | |
| Output A | 15 | 25 |
| Output B | 10 | 6 |
| Total | | 31 |
| **Year 3** | | |
| Output A | 20 | 28 |
| Output B | 10 | 8 |
| Total | | 36 |

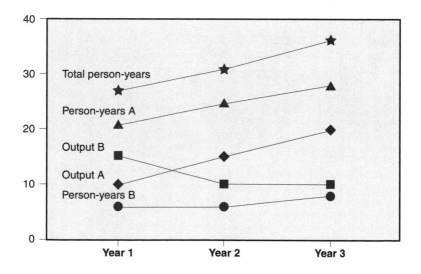

**Figure 6.4**   Basic data for calculating weighted productivity indexes over time.

costs, or unit labor requirements. When ULR weights are used, the different outputs can be weighted by the relative unit labor requirements of the standard organization or of the base year. The calculation of ULR weights is shown in Example 6.5 and illustrated in Figure 6.5.

The unit labor requirement measure is simply the reciprocal of the productivity measure. Productivity is measured as output per input. The unit labor requirement is measured as input per unit of output. The ULR measures how many units of labor time, for example, person-years, person-hours,

---

### EXAMPLE 6.5

### Calculating Unit Labor Requirement (ULR) Weights

|          | Output year 1 | Input year 1 | Unit labor requirements |
|----------|---------------|--------------|-------------------------|
| Output A | 10            | 21           | 21/10 = 2.1             |
| Output B | 15            | 6            | 6/15 = 0.4              |

---

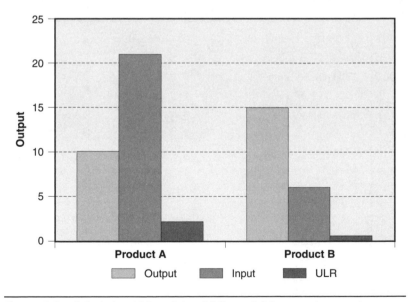

**Figure 6.5**   Calculating ULR weights (ULR = unit labor requirement in year 1).

or person-minutes, are required to produce one unit of output. It is usually expressed by the ratio of person-time per unit of output. The ULR measure tends to be more convenient to use when one unit of output requires a number of person-hours, or a certain fraction of person-hours, to produce it. As, however, this is only a straight reciprocal of the usual labor productivity measure, it will not be treated further in our discussion.

Returning to the discussion of weighting, we have said that the weighting should be done by an appropriate and measurable relative characteristic of the compared units, for instance by unit costs (that is, cost of each unit) or unit labor requirements, (that is, the labor input required to produce one unit of output). As the comparison is made with a standard or base year, the weights should be the unit costs, or unit labor requirements of the standard organization in the base year.

---

### EXAMPLE 6.6

### Calculating Weighted Output Indexes over Time

|  | Output | Base year ULR weights | Weighted output | Output index |
|---|---|---|---|---|
| **Year 1 (Base)** | | | | |
| A | 10 | 2.1 | 21 | |
| B | 15 | .4 | 6 | |
| Total | | | 27 | 100 |
| **Year 2** | | | | |
| A | 15 | 2.1 | 31.5 | |
| B | 10 | .4 | 4 | |
| Total | | | 35.5 | 131.5* |
| **Year 3** | | | | |
| A | 20 | 2.1 | 42 | |
| B | 10 | .4 | 4 | |
| Total | | | 46 | 170.4† |

Note: The output weighted by base year unit labor requirements means how many person-years would be required to produce the outputs of the current year if the base year level of productivity remained unchanged.

* 35.5:27 × 100 = 131.5
† 46:27 × 100 = 170.4

The figures have been rounded.

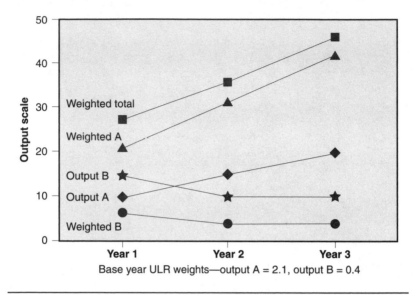

**Figure 6.6**    Calculating weighted output indexes over time.

As the productivity measures used are mostly output per person or output per person-hour, the unit labor requirements are the preferable weights. This is particularly so in offices or in public-sector operations where labor input tends to be the most significant input. When unit labor requirement weights are not available, unit value-added weights or unit labor cost weights are among the possible alternative weights.

## Calculating Weighted Output Indexes over Time

The construction of weighted productivity indexes over time follows the same principles and methods described. The quantity of the output of each product or service produced in each year is weighted by the labor required (ULR) to produce one unit of output in the base year. The procedure used is shown in Example 6.6, and illustrated by Figure 6.6.

## Calculation of Input Indexes

The input indexes are developed by dividing each year's person-years or person-hours by the person-years or person-hours of the base year, respectively, and multiplying by 100. Person-years or person-hours are considered additive, without weighting by occupation, training, or salary. The calculation of input indexes is shown in Example 6.7 and illustrated by Figure 6.7.

### EXAMPLE 6.7

### Calculating of Input Indexes over Time

| Year | Total person-years | Index |
|------|--------------------|-------|
| 1 | 27 | 100 |
| 2 | 31 | 114.8* |
| 3 | 36 | 133.3† |

\* 31:27 × 100 = 114.8
† 36:27 × 100 = 133.3

The data have been rounded.

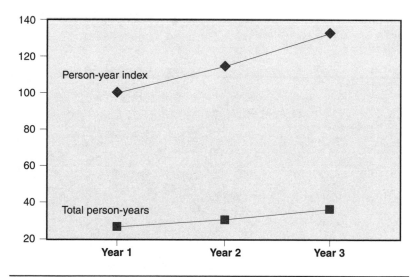

**Figure 6.7**   Calculation of input indexes over time.

## Calculating Productivity Indexes from Weighted Output and Input Indexes

The weighted output indexes are divided by the person-year (or person-hour) indexes, and multiplied by 100 to yield the productivity indexes, shown in Example 6.8, and illustrated by Figure 6.8.

## EXAMPLE 6.8

### Calculating Productivity Indexes from Weighted Output and Input Indexes

| Year | Output index (a) | Person-year (or person-hour) index (b) | Productivity index (output per person-year or per person-hour) (c) |
|---|---|---|---|
| 1 | 100 | 100 | 100 |
| 2 | 131.5 | 114.8 | 114.5* |
| 3 | 170.4 | 133.3 | 127.8† |

\* 131.5:114.8 × 100 = 114.5
† 170.4:133.3 × 100 = 127.8

The data have been rounded.

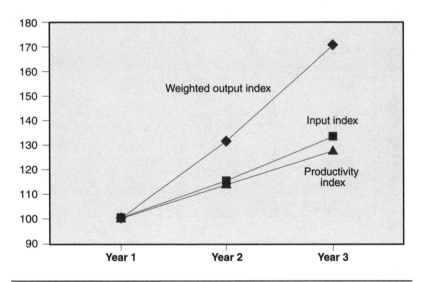

**Figure 6.8** Calculating productivity indexes from weighted output and input indexes.

If the data are in dollars rather than volume figures, the price changes due to inflationary influences need to be eliminated. This is obtained by dividing the index of the output expressed in dollar terms by the appropriate price index before the productivity index is calculated.

If the quality of output changes, it also should be reflected in the output measures. However, in view of the difficulty or even impossibility of quantifying quality changes that affect the value of the product or service only because of taste or other value judgements, it is practical to measure quality changes only if the base year input requirement of the higher-quality product or service would have been higher than for the old product. In such cases one customary procedure is to weight the higher-quality product or service by its estimated base year unit labor requirement, that is, how much more labor would have been required in the base year to produce the better product or service than the old product or service.

## Comparison of the Productivity of Two Organizations with Heterogeneous Products

If the productivity of two organizations with heterogeneous products is compared at one moment in time, the procedure can be done on the basis of the same principles, as shown in Example 6.9 and illustrated by Figure 6.9.

***Explanation of Example 6.9.*** The productivity of word processing letters (20/4 = 5) is five letters per word-processor-hour in organization A, and (21/7 = 3) three letters per word-processor-hour in organization B. Comparing productivity, the calculation is: 5/3 times 100 = 166.7. Therefore, organization A is 66.7 percent more productive than organization B in producing letters. In the productivity of handling phone calls, organization B is, however, better. The productivity comparison calculation shows that organization B is 10 percent better than organization A in handling telephone calls. (33/30 times 100 = 110, that is, 10 percent better).

The ULR of the letters word-processed by organization B in the base year was 7/21 = 0.333 and of phone calls, also of the standard organization B in the base year, was 6/198 = 0.03. When the output of both letters and phone calls of both organizations (columns a and d) are weighted (multiplied) by the unit labor requirements of each product in the standard organization B, we obtain the weighted outputs of both products in both organizations, so that they can be added up, namely, 6.7 and 1.82, yielding a weighted output figure of 8.52 for organization A, and 7 and 6 yielding a weighted output of 13 for organization B. When these weighted combined outputs are then divided by the respective total inputs (columns b and e) the combined weighted productivity figures are produced in columns c and f (1.42 and 1).

It is interesting to observe the importance of detail. In the combined operation, organization A is shown 42 percent more productive than organization B (1.42/1 times 100 = 142 − 100 = 42%). This overall result is hiding the fact, however, that while the productivity of organization A, as we have

## EXAMPLE 6.9

### Comparing the Productivity of Two Organizations with Heterogeneous Products Using Unit Labor Requirement (ULR) Weights

| | Organization A | | | Organization B | | |
|---|---|---|---|---|---|---|
| | Output (a) | Input hours (b) | Productivity (c) | Output (d) | Input hours (e) | Productivity (f) |
| Letters processed, no. | 20 | 4 | 5 | 21 | 7 | 3 |
| ULR/letter, hr. | 0.333 | | | 0.333 | | |
| Weighted | 6.7 | | | 7 | | |
| Phone calls, no. | 60 | 2 | 30 | 198 | 6 | 33 |
| ULR/call, hr. | 0.03 | | | 0.03 | | |
| Weighted | 1.82 | | | 6 | | |
| Total weighted both products | 8.52 | 6 | 1.42* | 13 | 13 | 1† |

Note: The quantity of each output is weighted (multiplied) by its ULR weight. The weighted outputs are then added up and divided by the inputs to yield their relative productivity measure, in this case 1.42 for organization A and 1 for organization B.

\* 8.52:6 × 100 = 1.42
† 13:13 × 100 = 1

The data have been rounded.

seen, is 66.7 percent better than organization B in the word processing of letters, it is about nine percent less productive in handling phone calls (30/33 × 100 = 91 – 100 = –9). Organization A will need to improve its telephone operation and organization B should concentrate on improving the handling of its word processing of letters.

In Chapter 6, the "Questions" and "Suggested Answers" that conclude the other chapters for practicing purposes are replaced by the following Task 6.1 to Task 6.9 tables, with spaces for practicing purposes, and corresponding Suggested Solution 6.1 to Suggested Solution 6.9 tables, which show the correct answers.

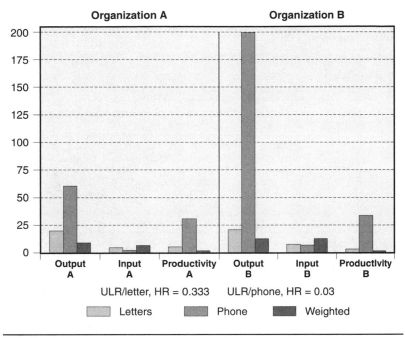

**Figure 6.9**    Productivity of two producers—heterogeneous outputs, ULR weights.

# TASKS FOR PRACTICING PRODUCTIVITY CALCULATIONS

## Task 6.1

Calculate the productivity data and the labor productivity index from the given figures. The correct results are shown as Solution 6.1.

**Calculation of productivity differences in percentages based on absolute data, when outputs are homogeneous.**

|  | Number of patients treated (a) | Number of nurse-hours (b) | Productivity (Output per nurse-hour) (c) | Labor productivity index (d) |
|---|---|---|---|---|
| Organization A | 360 | 24 | _____ | _____ |
| Organization B | 360 | 20 | _____ | _____ |

Therefore, the productivity of organization B is _____ % _____ than that of organization B.

## Solution 6.1

**Calculation of productivity differences in percentages based on absolute data, when outputs are homogeneous.**

|  | Number of patients treated (a) | Number of nurse-hours (b) | Productivity (Output per nurse-hour) (c) | Labor productivity index (d) |
|---|---|---|---|---|
| Organization A | 360 | 24 | 15 | 100 |
| Organization B | 360 | 20 | 18 | 120 |

Therefore, the productivity of organization B is 20% higher than that of Organization B.

Explanation: $360:24 = 15$; $360:20 = 18$; $18 \times 100:15 = 120$; $120 - 100 = 20$.

## Task 6.2

Calculate productivity indexes from raw data and calculate productivity over time when outputs are homogeneous.

**Calculation of productivity indexes from raw data and calculation of productivity changes over time, when outputs are homogeneous**

| Year | Number of patients treated | Number of nurse-hours | Number of patients nurse-hours | Labor productivity index (Year 1 = 100) | Productivity change from previous year % |
|---|---|---|---|---|---|
| 1 | 360 | 24 | _____ | _____ | _____ |
| 2 | 480 | 30 | _____ | _____ | _____ |
| 3 | 504 | 28 | _____ | _____ | _____ |
| and so on | | | | | |

## Solution 6.2

**Calculation of productivity indexes from raw data and calculation of productivity changes over time, when outputs are homogeneous**

| Year | Number of patients treated | Number of nurse-hours | Number of patients nurse-hours | Labor productivity index (Year 1 = 100) | Productivity change from previous year % |
|---|---|---|---|---|---|
| 1 | 360 | 24 | 15 | 100.0 | 0 |
| 2 | 480 | 30 | 16 | 106.7 | 6.7 |
| 3 | 504 | 28 | 18 | 120.0 | 12.5 |

Explanation: 360:24 = 15; base year index = 100%; 480:30 = 16; 16:15 × 100 = 106.7%; change from base (= previous) year = 6.7%

## Task 6.3

Construct productivity indexes from output and input indexes, when outputs and inputs are homogeneous. The correct results are shown as Solution 6.3.

**Construction of productivity indexes from output and input indexes, when outputs are homogeneous**

| Year | Number of patients | Output converted into index (Year 1 = 100) | Input of nurse-hours | Input converted into index (Year 1 = 100) | Productivity index (Year 1 = 100) |
|---|---|---|---|---|---|
| 1 | 360 | _____ | 24 | _____ | _____ |
| 2 | 480 | _____ | 30 | _____ | _____ |
| 3 | 504 | _____ | 28 | _____ | _____ |

and so on

## Solution 6.3

**Construction of productivity indexes from output and input indexes, when outputs are homogeneous**

| Year | Number of patients | Output converted into index (Year 1 = 100) | Input of nurse-hours | Input converted into index (Year 1 = 100) | Productivity index (Year 1 = 100) |
|------|------|------|------|------|------|
| 1 | 360 | 100.0 | 24 | 100.0 | 100.0 |
| 2 | 480 | 133.3 | 30 | 125.0 | 106.6 |
| 3 | 504 | 140.0 | 28 | 116.7 | 120.0 |

Explanation: All base year indexes are 100.0.
$480 \times 100:360 = 133.3$; $30 \times 100:24 = 125.0$; $133.3 \times 100:125.0 = 106.6$
$504 \times 100:360 = 140.0$; $28 \times 100:24 = 116.7$; $140.0 \times 100:116.7 = 120.0$

## Task 6.4

Basic data for calculating weighted productivity indexes over time. Use the following basic data to calculate the unit labor requirement (ULR) weights in Task 6.5. You will be asked in Task 6.6 to do the final calculation of weighted productivity indexes over time, using the basic data from Task 6.4 and the results from Task 6.5.

**Basic data for calculating weighted productivity indexes over time**

|  | Number of patients | Nurse-hours |
|------|------|------|
| **Year 1** |  |  |
| Treatment A | 360 | 24 |
| Treatment B | 96 | 16 |
| Total |  | 40 |
| **Year 2** |  |  |
| Treatment A | 360 | 22 |
| Treatment B | 140 | 22 |
| Total |  | 44 |
| **Year 3** |  |  |
| Treatment A | 396 | 22 |
| Treatment B | 160 | 20 |
| Total |  | 42 |

## Solution 6.4

Use the following basic data for calculating weighted productivity indexes over time. For your convenience, these are repeated from Task 6.4. The correct results are shown in Solutions 6.6 through 6.9.

**Basic data for calculating weighted productivity indexes over time**

|  | Number of patients | Nurse-hours |
|---|---|---|
| **Year 1** | | |
| Treatment A | 360 | 24 |
| Treatment B | 96 | 16 |
| Total | | 40 |
| **Year 2** | | |
| Treatment A | 360 | 22 |
| Treatment B | 140 | 22 |
| Total | | 44 |
| **Year 3** | | |
| Treatment A | 396 | 22 |
| Treatment B | 160 | 20 |
| Total | | 42 |

## Task 6.5

Using the basic data that have been transferred from Task 6.4, calculate the base year unit labor requirement weights that will be needed in the tasks below.

**Calculating unit labor requirement (ULR) weights**

|  | Number of patients Year 1 | Nurse-hours and minutes Year 1 | ULR weights in base year = Year 1 |
|---|---|---|---|
| Treatment A | 360 | 24 _____ | _____ |
| Treatment B | 96 | 16 _____ | _____ |

## Solution 6.5

Using the basic data that have been repeated in Solution 6.4, calculate the base year unit labor requirement weights that will be needed in the tasks below.

**Calculating unit labor requirement (ULR) weights**

|  | Number of patients Year 1 | Nurse-hours and minutes Year 1 | ULR weights in base year = Year 1 |
|---|---|---|---|
| Treatment A | 360 | 24 × 60 = 1440 | 1440:360 = 4 |
| Treatment B | 96 | 16 × 60 = 960 | 960:96 = 10 |

Explanation: 24 hours multiplied by 60 in order to be expressed in minutes = 1440. 1440 minutes divided by 360 patients = 4 minutes per patient. Similarly, 16 × 60 = 960; 960:96 = 10.

## Task 6.6

Using the data given in Task 6.4 and calculated in Solution 6.5, calculate the weighted output indexes over time.

**Calculating weighted output indexes over time**

|  | Number of patients | ULR weights in base year | Weighted output | Weighted output index (Base year = 100) |
|---|---|---|---|---|
| **Year 1** |  |  |  |  |
| Treatment A | 360 | 4 | _____ |  |
| Treatment B | 96 | 10 | _____ |  |
| Total |  |  | _____ | _____ |
| **Year 2** |  |  |  |  |
| Treatment A | 360 | 4 | _____ |  |
| Treatment B | 140 | 10 | _____ |  |
| Total |  |  | _____ | _____ |
| **Year 3** |  |  |  |  |
| Treatment A | 396 | 4 | _____ |  |
| Treatment B | 160 | 10 | _____ |  |
| Total |  |  | _____ | _____ |

## Solution 6.6

**Calculating weighted output indexes over time**

| | Number of patients | ULR weights in base year | Weighted output | Weighted output index (Base year = 100) |
|---|---|---|---|---|
| **Year 1** | | | | |
| Treatment A | 360 | 4 | 1440 | |
| Treatment B | 96 | 10 | 960 | |
| Total | | | 2400 | 100.0 |
| **Year 2** | | | | |
| Treatment A | 360 | 4 | 1440 | |
| Treatment B | 140 | 10 | 1400 | |
| Total | | | 2840 | 118.3 |
| **Year 3** | | | | |
| Treatment A | 396 | 4 | 1584 | |
| Treatment B | 160 | 10 | 1600 | |
| Total | | | 3184 | 132.7 |

Explanation: Year 1: $360 \times 4 = 1440$; $1440 + (96 \times 10) = 2400 =$ the base year index of 100.0. Year 2: $360 \times 4 = 1440$; $140 \times 10 = 1400$; $1440 + 1400 = 2840$; productivity index = 118.3. Year 3: $396 \times 4 = 1584$; $160 \times 10 = 1600$; $1584 + 1600 = 3184$; productivity index = 132.7.

## Task 6.7

Using the data given in Task 6.4, calculate the input indexes.

**Calculating input indexes**

|  | Number of nurse-years (or nurse-hours) | Input indexes (Year 1 = 100) |
|---|---|---|
| **Year 1** | | |
| Treatment A | 24 | |
| Treatment B | 16 | |
| Total | _____ | _____ |
| **Year 2** | | |
| Treatment A | 22 | |
| Treatment B | 22 | |
| Total | _____ | _____ |
| **Year 3** | | |
| Treatment A | 22 | |
| Treatment B | 20 | |
| Total | _____ | _____ |

## Solution 6.7

**Calculating input indexes**

|  | Number of nurse-years (or nurse-hours) | Input indexes (Year 1 = 100) |
|---|---|---|
| Year 1 | | |
| Treatment A | 24 | |
| Treatment B | 16 | |
| Total | 40 | 100.0 |
| Year 2 | | |
| Treatment A | 22 | |
| Treatment B | 22 | |
| Total | 44 | 110.0 |
| Year 3 | | |
| Treatment A | 22 | |
| Treatment B | 20 | |
| Total | 42 | 105.0 |

Explanation: Labor input is usually considered homogeneous in productivity measures and is therefore unweighted.

## Task 6.8

Using the output and input indexes calculated in Solution 6.6 and Solution 6.7, calculate the weighted output per nurse-year productivity indexes.

**Calculating productivity indexes from (weighted) output and input indexes**

| Year | Weighted output index (Year 1 = 100) | Nurse-year (input) index (Year 1 = 100) | Productivity index (weighted output index: nurse-year index × 100) (Year 1 = 100) |
|---|---|---|---|
| 1 | 100.0 | 100.0 | _____ |
| 2 | 118.3 | 110.0 | _____ |
| 3 | 132.7 | 105.0 | _____ |

## Solution 6.8

**Calculating productivity indexes from (weighted) output and input indexes**

| Year | Weighted output index (Year 1 = 100) | Nurse-year (input) index (Year 1 = 100) | Productivity index (weighted output index: nurse-year index × 100) (Year 1 = 100) |
|---|---|---|---|
| 1 | 100.0 | 100.0 | 100.0 |
| 2 | 118.3 | 110.0 | 107.5 |
| 3 | 132.7 | 105.0 | 126.4 |

Explanation: Year 1 = 100.0; Year 2: 118.3:110 × 100 = 107.5; Year 3: 132.7:105.0 × 100 = 126.4. The figures have been rounded.

## Task 6.9

Calculate and compare the productivity of two organizations with heterogeneous products, using ULR weights. The correct results are shown as Solution 6.9.

**Comparing the productivity of two organizations with heterogeneous products, using ULR weights**

|  | Nursing unit 1 | | | Nursing unit 2 | | |
| --- | --- | --- | --- | --- | --- | --- |
|  | Output (a) | Nurse-hours (b) | Productivity (c) | Output (d) | Nurse-hours (e) | Productivity (f) |
| Diagnostic A | 360 | 24 | ____ | 360 | 20 | ____ |
| ULR in diagnostic A | 4 | | | ____ | | |
| Weighted | ____ | | | ____ | | |
| Diagnostic B | 96 | 16 | ____ | 175 | 35 | ____ |
| ULR in diagnostic B | 10 | | | ____ | | |
| Weighted | ____ | | | ____ | | |
| Total weighted both diagnostic treatments | ____ | 40 | ____ | ____ | 55 | ____ |

## Solution 6.9

**Comparing the productivity of two organizations with heterogeneous products, using ULR weights**

|  | Nursing unit 1 | | | Nursing unit 2 | | |
| --- | --- | --- | --- | --- | --- | --- |
|  | Output (a) | Nurse-hours (b) | Productivity (c) | Output (d) | Nurse-hours (e) | Productivity (f) |
| Diagnostic A | 360 | 24 | 15 | 360 | 20 | 18 |
| ULR in diagnostic A | 4 | | | 4 | | |
| Weighted | 1440 | | | 1440 | | |
| Diagnostic B | 96 | 16 | 6 | 175 | 35 | 5 |
| ULR in diagnostic B | 10 | | | 10 | | |
| Weighted | 960 | | | 1750 | | |
| Total weighted both diagnostic treatments | 2400 | 40 | 60 | 3190 | 55 | 58 |

Explanation: 360:24 = 15; 360:20 = 18; from Solution 5, the ULR in diagnostic A = 4; weighted (360 × 4) = 1440 minutes; 96:16 = 6; 175:35 = 5; ULR in diagnostic B = 10; weighted (96 × 10) = 960; 175 × 10 = 1750; total weighted in unit 1 = 1440 + 960 = 2400; 2400:40 = 60; in 3190:55 = 58 in unit 2 = 1440 + 1750 = 3190.

Therefore, although the overall productivity is a little higher in nursing unit 1 (60 × 100:55 = 103.4), this is only due to its 20 percent higher productivity (5 × 100:6 = 120) in diagnostic work "B." It could improve its overall productivity significantly by focusing on improving diagnostic work "A" in which nursing unit 2 has 20 percent (18 × 100:15 = 120) better results. Nursing unit 2 could, however, improve its overall productivity by concentrating on improving diagnostic work.

# 7

# Benefits of Benchmarking

## THE IMPORTANCE OF COMPARATIVE ORGANIZATIONAL PERFORMANCE ANALYSIS

We have already noted that "productivity" is meaningful only if compared with some benchmark or standard. Is performance more or less productive than that of others, say, competitors? We have also pointed out that productivity can be affected primarily at the organizational level, that is, that of companies, their component units, or organizational units in public-sector organizations.

It also has been pointed out that the importance of microeconomic productivity, that is, productivity at the plant level, and its factors had been recognized for a long time. However, it was not until after World War I that the importance of organizational productivity was recognized when the RKW (Rationalisierungs Kuratorium der Deutschen Wirtschaft) was set up in Germany with the objective of "rationalizing," that is, improving, the efficiency of companies. After World War II the United States Department of Labor initiated a company-level productivity measurement and analysis program related to the Marshall Plan.

Having realized the vital importance of comparative performance measurement and improvement at the level of the company or other organizational unit, the author of this book developed an integrated productivity and profitability interfirm comparison (IFC) program in about 1970. The program was later extended to international comparisons with groups of companies in the United States and Europe, as well as to Asian countries.

These comparisons went far beyond the conventional profitability comparisons, which are affected by many financial and other economic influences. This system highlighted the productivity factors and provided for the adjustments that are necessary for revealing the factors that influence productivity. While profitability is the immediate objective of private-sector organizations, its long-term determinant is productivity, how much real output is produced from the real resources used.

The interfirm profitability and productivity comparisons had been built with the principles of productivity measurement and analysis, that is, the data of the participating companies are adjusted for comparability with each other. Company managers who are preparing their operating plans need reliable measures of the best capital/labor ratios in order to ensure the most efficient capital intensity. Of course, care must be taken to allow for differences in the basket of goods produced or services provided, processes used, degree of processing, and markets served. The lessons learned, which are described below, are useful and proven guidelines for any organization that wants to compare itself with any other benchmark organization.

In Canada alone, more than 4000 companies have participated in interfirm comparisons in over 100 industries. Several comparisons were conducted on an international basis, with the United States and several European countries. The author also conducted productivity interfirm comparisons between Asian countries.[1,2]

The interfirm productivity and profitability comparison program in North America was started in order to stimulate productivity growth. After the introductory period, it was left to be carried out by the private sector.

Since the importance of productivity improvement is not generally well understood—although it is always talked about—public assistance was highly desirable. However, it can be done on a private-sector basis as has been done for several decades in Great Britain and around the world by the Centre for InterFirm Comparison.[3] (That program had originally been initiated by the British Institute of Management.) It can also be organized by the appropriate industry association for the benefit of its members.

The procedure used in these comparisons is briefly described in the following. The fixed assets are revalued on a comparable basis. The objective is to ensure a reasonable measure of return on investment. As it is usual to express profits as a percentage of the original investment, one can do so, but must convert the original cost to "constant dollars," that is, what the original cost would be if valued in today's dollars. Official price indexes are used for machinery, equipment, and buildings in today's dollars. The fiscal years of the participating companies are brought in line with each other. The original book values of capital purchased each year during the true economic life of the assets are collected and converted to the current year's values. The accuracy of the recalculated values of the assets can be checked against replacement values and current new equipment values.

The correct data of the true length of economic life of assets are obviously important factors of productivity. Taxation data are defined in accordance with political considerations and are usually not suitable for productivity measurement purposes because, for instance, of accelerated depreciation, and so on. The acceptable data need to be obtained through discussion with leading member companies in the particular industry or sector. The true economic life of the various types of assets needs to be used for amortization purposes. The average age of capital equipment is also very important, not only because of its impact on equipment maintenance but also because of the varying degree of embodied new technology.

As it concerns the productivity of the organization, it does not matter whether the fixed assets are owned or leased. Both are inputs. The conversion factors from leased to owned assets or vice versa can usually be obtained from leasing companies. Then the assets need to be depreciated using a standard straight-line rate, but over their true economic life rather than at the rate prescribed for taxation purposes.

It is important to decide whether "available" or "used" capital assets should be included in the measures. It is preferable to use the "used" fixed assets as capital inputs because unused assets are not contributing to the output. The only unused assets that should be included in capital input are those assets that are retained for use at peak periods to serve as reserve capacity. Unused land should only be included as inputs if it is considered required for the conduct of the organization's current operations rather than speculation.

Working capital consists of inventories and accounts receivable. Neither of these should be used in calculating productivity measures. Inventories consist of new materials that have not been used yet in the production process as inputs, or goods in process and finished goods that have already been or will eventually be counted in the output.

The problems faced in the productivity analysis include the valuation of sales with or without discounts. Adjustments may also be needed to ensure comparability in the:

- Person-hours worked (as opposed to person-hours paid)

- Valuation of fixed assets built by own labor

- Production costs in materials, labor, and other inputs

- Production cost of goods sold

- Purchases of secondhand equipment, or older equipment traded for newer equipment

- Purchases made in instances of forced sale

- Government subsidies for new equipment

- Differences in valuation practices among compared companies

- Warehousing, distribution, shipping, promotion, and marketing costs

- The inclusion of freight paid on outgoing goods in sales figures

- What to include in the floor area

# THE METHOD OF INTEGRATED ORGANIZATIONAL PERFORMANCE ANALYSIS

In this analytical system the profitability and productivity ratios are organized in the form of an integrated set of ratios. All the measures used in the ratios need to be developed according to the productivity principles with respect to deflation, depreciation, and so on. Then "pure" productivity measures are to be included to complete the system, for example, output per hour of work, number of hotel rooms cleaned by a maid.

*Operating profit* means the income earned in employing the operating assets, calculated after depreciation at standardized rates but before deducting interest charges and before taxes. It represents the total earnings of all operating assets without regard to how they are financed. *Operating assets* means the assets actually employed in the production process of the enterprise, both fixed and current, excluding such items as marketable securities, investment in other enterprises, goodwill, or other intangibles.

The term used in the integrated performance analysis of *sales value of production* is not generally known. It expresses the year's production in term of sales dollars (or whatever currency is applicable). Sales value of production consists of all manufacturing (or service) costs less adjustments for changes in inventories of finished goods and goods in process. It is calculated by multiplying production costs by the ratio of sales to cost of goods sold. "Sales value of production" equals "sales" if production costs equal cost of goods sold, that is, when there is no change in inventories. (This adjustment is necessary because not all the goods sold in this year were necessarily produced with this year's resources and, vice versa, some of the goods produced with this year's resources may only be sold next year. The inputs and outputs must match!)

The interfirm comparison program must be conducted with guaranteed confidentiality. The results are conveyed to the participating companies only, and each firm is told only which set of data is theirs. The names of all companies are replaced by code letters. The participating group of companies, if they all agree, are given the confidential data (without company names), ranges, and averages, in order to show them how they compare with their competitors. It is interesting to note that even the best performers have almost invariably found areas in which they can learn from the comparison.

Productivity measures are of little use unless their findings are followed up by corrective action. To enable action, the productivity measures need to be interpreted and their significance and implications identified. The producers of the productivity measures can not expect that the clients or users are familiar with productivity concepts and techniques. The interpretation must be clear and specific. It should not be mechanical because a high or low figure is not necessarily good or bad. It has to be examined in view of the environment and with regard to the interrelationships with the various input and output factors.

The best is to proceed step by step. One should examine the profitability ratios and the pure productivity ratios as well as their interrelationships. The variations, differences, changes in the results need to be examined, both in total and among the components. Are they better or worse than those of other comparable units? Why? Did they get better or worse, and why? One should ask which elements show relatively high variations or changes. It needs to be determined whether it was the changes in outputs or in inputs that could have caused the changes or variations in productivity. One should examine in perspective what the findings are and how they relate to each other and to the total. The results should be compared with the base. The possible factors and causes should be eliminated one by one until as much as possible of the changes or variations is explained.

Then one should concentrate on a few main findings. It needs to be explained why the measures led to this or that conclusion. The findings must emanate from the measures. Preconceived notions that are not supported by the measures must not be accepted. Conclusions need to be reached as to what actions seem to be warranted, and what will be their likely costs and results. The productivity measures need to be integrated, if possible, with the existing financial management information system. This is important because the productivity measures have financial and budgetary implications. Also, the productivity measures will be taken more seriously if they form a part of the overall management information system, rather than stand aside by themselves.

The following financial ratios are the most commonly used by enterprises. The ratios shown in italic letters (numbers 4, 5, 6, 8, and 11) are closely related to productivity. However, for productivity measurement purposes they need the appropriate adjustments. Most of the 15 performance ratios frequently gathered by companies are financial performance ratios. All these ratios are, however, oriented toward specific aspects of financial performance rather than managerial productivity performance, which would indicate strong or weak areas of human or physical resource utilization in the organization. The existing ratios are prepared for shareholders or for taxation purposes, with their respective interests or rules in mind:

1. Fixed assets to equity

2. Long-term debt to equity

3. Interest to total cost

4. *Cost of sales to sales*

5. *Sales to inventory*

6. *Collection period (in days)*

7. Current assets to current liabilities

8. *Profits before tax on capital employed*

9. Profits before tax on equity

10. Profits before tax on total income

11. *Profits before tax on total income excluding investment income*

12. Cash earnings on sales

13. Profits after tax on capital employed

14. Profits after tax on equity

15. Profits after tax on total income

At company-level or public-sector organizational-unit level, productivity measures are not often used as yet, mainly because management tends to be unaware of the usefulness of this modern tool of management and is not familiar with the methodology of productivity measurement. Progress is, however, being made. Industry and trade associations gather various data on their members' operations. Some of these are, in fact, productivity ratios, such as the number of tables served per waiter or waitress in the restaurant industry, or square meters of floor space per value of sales in retail trade. Nevertheless, these tend to be independent, specific measures, not integrated into an overall performance evaluation.

The enterprise-level performance analysis program has proven that productivity analysis can identify very significant performance improvement problems and potential in virtually every enterprise if the productivity and profitability measures are built into an integrated framework. The managers of all enterprises that participated in our comparisons have concluded that the combined profitability and productivity analysis provides them with important new insights into their operations. These data reveal the company's specific problems and opportunities and are of great practical value to the managers.

Experience has shown that in order to be of interest to enterprise management, the ratios have to concentrate on the profitability of the enterprise, that is, the return on investment, which is of direct interest to managers. The pure productivity ratios need to be added to those. All ratios must be easily understood. The most useful information varies from industry to industry because, for instance, in labor-intensive industries more information is needed on labor requirements, while in capital-intensive industries more detail is wanted on the capital available and used.

The organization's strengths and weaknesses need to be identified in comparison with a standard. The question is whether the various ratios show strengths or weaknesses compared to the standard. The standard can be a comparable organization or group of comparable organizations at the same moment in time, or the same organization at the base period. Does the organization show better or worse performance in respect to specific ratios than the standard organization or is it getting better or worse? If so, why?

In integrated profitability and productivity analysis usually some 30 to 40 ratios are highlighted for an organization to reflect such typical operating ratios as the costs of labor, materials, production overhead, selling, administration as percentage of sales, and a set of supplementary ratios designed more specifically to indicate productivity. The latter could be value added per person-year, per person-hour, or per unit of raw material, the number of products produced per dollar of machinery and equipment, or the number of dress shoes produced per square meter of leather.

In addition to the basic set of data, many more quantitative and qualitative data are gathered in order to help the analyst explain the impact of the variations and changes on the organization's overall profitability and productivity. Various ratios also can be usefully subdivided into components. For example, "direct production labor cost/sales value of production" in restaurants can be subdivided into direct labor in food *production* and direct labor in food *service* (waiters and waitresses), as well as in room service.

All the ratios are constructed as an integrated set, starting from the primary ratio, the rate of return on operating assets, which is of primary importance to enterprises. This ratio measures the earning power of the enterprise and indicates whether management is using its resources effectively. The analysis is extended on one side to the various costs per sales value that determine the profit margin, and on the other side to the various categories of assets in proportion to sales, for example, buildings, equipment, inventories, accounts receivable, and, where meaningful and significant, land as well.

In order to improve comparability, a number of adjustments are needed. These can not be examined here in detail but include finding solutions to such problems as variations or changes in the basket of goods or services produced, production costs in materials, labor and other inputs, warehousing, distribution, and shipping, labor hours and skill composition, and allocation of central costs (for example, accounting, security) to such various cost centres as purchasing, production, maintenance, warehousing, distribution, and so on.

# PERFORMANCE IMPROVEMENT OPPORTUNITIES REVEALED BY BENCHMARKING

Worldwide experience with the above described integrated profitability and productivity analysis has indicated specific improvement potentials particularly in labor management, materials management, production management, administration, as well as in selling and promotion. The analyses have focused attention on the need for a better utilization of human, physical, and/or financial resources, better management of time, as well as the optimal choice of product variety. The comparisons provide participating enterprises with a meaningful productivity-oriented management information system through which they can systematically monitor their progress and/or compare themselves with an appropriate industry standard.

Corrective actions have ranged throughout the entire spectrum of business operations. Important opportunities have been found in the planning and forecasting of sales and their direct and indirect labor requirements, in the utilization of fixed assets, and the effective use of energy and consumable materials. As a result of interfirm comparison findings, several companies reported increases in shop size to allow greater volume of sales with little increase in overhead. Others developed new marketing strategies for increasing sales even if this demanded sacrifices in other departments.

It was found that inadequate analysis and planning led to labor imbalances. In order to reduce labor costs—which often are a major factor of total costs—some companies introduced a continuing audit of productivity with particular attention to the numbers employed in relation to the volume of goods produced. In order to minimize labor turnover, firms took steps to identify its causes. Changing the product mix often allowed improvements in shop labor productivity and the corresponding reduction of labor costs.

Production outputs have been improved by the reduction of product variety through fewer models and procedures. The interfirm comparisons showed that the generally assumed desirability of increasing scale was unfounded. It was true for some industries but not for others. What was found was that performance benefits can be derived usually from product-specific increases of scale in the form of increased lengths of production runs.

One company learned that it had a very high hourly labor rate but that this was offset by their high labor productivity, and that their relatively high labor rate has not impaired their competitiveness. Their main problem was low capacity utilization and overly wide market area that resulted in too high distribution costs. The interfirm comparisons also revealed productivity and profitability opportunities in the areas of fixed assets, machinery utilization, increasing automation where justified, and in the selling of unproductive assets. In the area of current assets, many companies took steps to improve the turnover of their working capital by adjusting their inventories. Many firms revised their billing and collection systems and strengthened the control of accounts receivable.

Several companies established systems to indicate the causes of excessive material costs. This led to tighter control of purchasing and a closer look at the prices paid for outside work. Other companies have identified productivity improvement opportunities from warehousing, shipping, and selling expense ratios. Improvement opportunities have also been indicated in distribution operations, in the utilization of their transport fleet, and traffic routing.

Some companies found that they had unreasonably high overhead costs that led to only average return on assets. The high overhead costs may have been due to high maintenance, repair, and energy costs, which needed to be better controlled. It was also indicated that if some of the companies could reduce their manufacturing overhead costs to the median level of the firms compared, their return on assets would rank high among their competitors. Some firms showed comparatively high labor costs in spite of above-average labor productivity. One of the main problem areas was in nonproductive expenses, including high administration, shipping, and selling costs that were the highest in their group of companies.

# QUESTIONS

Q: 7-1   List some of the main cost elements of a manufacturing firm.

Q: 7-2   How do you establish the full cost of the various services you are providing?

Q: 7-3   Give some examples of data useful for productivity measurement in physical distribution.

Q: 7-4   What are some examples of productivity measures in hotels and motels?

Q: 7-5   Give examples of productivity measures in restaurants.

Q: 7-6   What does "value added" mean?

# SUGGESTED ANSWERS

A: 7-1   Some of the main cost elements of a manufacturing firm include:

- Costs of purchasing

- Materials and components

- Direct and indirect manufacturing labor costs

- Costs of fringe benefits

- Manufacturing overhead

- Operating expenses (management and administration, promotion and selling, warehousing and shipping)

- Service inputs

- Fixed assets

- Current assets

- Financing costs

A: 7-2   One can establish the full cost of each of the various services that are provided by:

- Measuring the time required by each activity and multiplying each by the total hourly compensation (including fringe benefits)

- Adding the relevant part of the fixed costs of premises and equipment, materials cost, and the relevant support cost

A: 7-3  Data that are useful for productivity measurement in physical distribution include accounting records, freight bills, and driver logs.

A: 7-4  Examples of productivity measures in hotels and motels include:

- Rooms sold per rooms available

- Number of guests per room

- Rooms sold per rooms departmental employees

- Total value of fixed assets per full-time equivalent employee

- Amount sold per dining room seat

- Covers sold per seat

- Number of available rooms per full-time equivalent employee

- Sales per employee (in various departments)

A: 7-5  Examples of productivity measures in restaurants include:

- Total sales per cover

- Segmented expenses per cover

- Sales per seat

- Segmented expenses per seat

- Sales per segmented expenses

A: 7-6  "Value added" means "net output," namely what value is "added" by the operation in question to the materials and components purchased. It can be expressed as the difference between gross output (sales) and purchases from outside the company or, alternatively, by adding up the cost of labor input, capital input, and the profit of the company. Both results should be identical.

# Part III

# Basic Elements of Productivity Improvement: Making Personal Efforts More Efficient and Effective

# 8

# Lessons from Successful Productivity and Quality Improvement Programs

## MANY BARRIERS TO PRODUCTIVITY HAVE BEEN OBSERVED

As people become better educated and new technologies, machines, and materials are invented, productivity—and the related work patterns and standard of living—are gradually improving. However, in today's world of rapidly increasing expectations and global competition, this "natural" productivity improvement is no longer sufficient to meet the needs. There are many productivity problems and opportunities, but taking actions to correct problems and taking advantage of the opportunities are far from automatic. The problems and opportunities—the impeding and impelling forces— must be identified and quantified, that is, measured, priorities must be set, and action taken on many fronts.

In systematically identifying and resolving productivity performance problems, it is usually best to start with the most obvious, very effective, and relatively easy approach of dealing constructively with counterproductivity. This is accomplished by identifying and eliminating, or at least minimizing, counterproductive forces that impede productivity in an organization. These could be barriers and/or disincentives.

The barriers stand in the way of productivity enhancement while the disincentives tend to discourage it. Sometimes it is difficult to distinguish productivity barriers from disincentives as they may overlap, depending on their exact nature and strength.

Typical examples of factors that could be barriers to productivity include the following:

- Resistance to change, being afraid of the unknown or of the assumed negative impact of change, such as loss of employment, power, or prestige. Some supervisors may resist change because they may worry that the change will interfere with their relationship with the workers. Resistance to change is often due to the lack of good communication. In the case of professional and other creative workers, the resistance to change may be caused because they feel that their output is difficult to quantify, their work is typically performed in groups that produce joint products or services, and because they often deliver their products internally, that is, to other units in the same organization. As such, their performance improvement may not "show" in a productivity measure.

- Apathy and/or negative attitude of management or staff resulting in a lack of attention to productivity.

- Widespread lack of knowledge or misunderstanding of what productivity is.

- Insufficient attention to productivity.

- The tendency in cases of increasing demand or emerging problems to spend more automatically without first considering whether a change in methods or procedures could resolve the problem.

- Short-range focus in budgeting that makes long-term improvement difficult.

- Budget cuts that are often made across the board with little regard for managerial efficiency in individual organizational elements, with the result that efficient elements are discarded together with the inefficient without differentiating between areas where reductions are needed or not justified.

- Poor organization.

- Bottlenecks that may be caused by poor planning due to which all operations lead to one assembly unit, by poorly maintained equipment that leads to frequent breakdowns and downtime, inadequate inventory planning and control, or unsatisfactory suppliers.

- Overinflated organization and/or staffing.

- Administrative constraints.

- Antiquated or incorrect job designs.

- Difficulty in measuring and analyzing productivity.
- Lack of sufficient data, and no or deficient management information systems.
- Lack of analytical ability.
- Lack of ability and skills to assimilate new technology.
- Lack of adequate communication and coordination among various segments of the organization, including insufficient awareness of the implications of actions of individual units on other parts of the organization, for example, engineering on administration or vice versa.
- Poor internal training policies and practices.
- Lack of clear understanding by workers of what is expected of them.
- Lack of effective appraisal and feedback of performance.
- The lack of appropriate productivity incentives, for instance, gainsharing or other rewards, is also a problem that is often reported as holding down productivity.
- Staff and industrial relations problems.
- Lack of sufficient personnel resources, which is due, for instance, to personnel ceilings.
- High management and/or employee turnover or absenteeism that, in turn, may have other underlying problems.
- Shortage of trained manpower also may prove to be a major barrier to productivity.
- Infrastructural problems, such as transportation or energy difficulties.
- Problems created by the clients, for example, changes in tastes, attitudes.
- Environmental problems such as constraints created by external bodies, including legal constraints.

Factors that tend to act mainly as disincentives to productivity include, among others:

- Inadequate remuneration levels.

- Ignored achievements, lack of recognition, taking good work for granted.

- Penalizing productivity savings.

Among the greatest problems facing management are the lack of knowledge of productivity measurement techniques and the lack of solid measurement data. As a result, managers and supervisors often fail to apply proper productivity measurement. The lack of meaningful comparable measurement data on both labor and capital productivity at the enterprise or office level, which could serve as standards of comparison at that level, is the other major measurement problem. Such data are sometimes available from interfirm comparisons or special surveys of industry associations, but the best ones tend to be considered confidential and limited to participants only. The lack of data for enterprise-level, office-level, or other unit-level productivity measurement is not limited to labor measures—particularly "hours worked"—but also includes, for instance, data on the age, structure, nature, cost, and length of economic life of fixed capital.

In the area of capital and technology, the most often reported problems occur in the following aspects:

- Inadequate supply of material or financial resources

- Inadequate use of capital; overcapitalization or undercapitalization

- Incorrect investment, misinvestment; equipment chosen without regard to needs

- Outdated equipment, lack of taking advantage of new technology where justifiable

- Inadequate use of personnel during equipment downtime

- Lack of balanced coordination of people, places, and machines

- Lack of training in the use of capital equipment

# THE PRODUCTIVITY IMPROVEMENT PROGRAMS OF THE PAST HAD AVOIDABLE WEAKNESSES

When productivity improvement programs are developed, it is important not to limit our efforts to eliminating or reducing problems, but also to enhancing favorable forces that could be strengthened further.

Such forces exist, for example, if there is high demand for the product or service provided, if there is a favorable government support policy or assistance from international organizations, when there are existing but underutilized production, training, or other facilities or programs, or even when improvement opportunities have been identified by criticism from the consumers of the products or services. Such opportunities should be sought out and utilized for improving productivity.

The objectives need to be defined, production targets set, and the actual performance compared to the targets. For example, the supervisors may compare the number of people or machines on a certain operation with how many of each they actually need. Efficiency differences can also be revealed by comparisons with similar organizations. The differences can be analyzed to determine the causes of the variations and to see whether the more efficient operations could be replicated in similar offices or other operations. Opportunities can often be spotted by endeavoring to remain responsive to clients, customers, or other users of the goods or services provided by the group.

Over the past few years, it has become clearly evident that organizations—large and small—that were successful in achieving significant productivity improvement had a number of common characteristics in their operations. These essential elements of successful productivity programs have been identified and will be reviewed in the following because it is important to learn from the successful accomplishments of others and to apply what we already know rather than try to reinvent the wheel.

The importance of people can't be overstated. It has already been mentioned that it is often assumed that increased productivity is brought about mainly by higher technology, larger scale of operations, or harder work. The scale of operations may be significant, particularly in certain industries or types of operations, and the good utilization of working time is essential. However, careful and systematic planning, proper organization, training, motivation, information sharing, staff involvement, incentives, and "smart" work are usually far more important determinants of productivity levels and growth than those so often assumed, such as more technology. By way of example, a study of banks found that banks with high employer profiles have one-third better return on assets and almost as much better return on equity than banks with less favorable employer attitudes.

The need for increasing productivity has led a growing number of organizations to start productivity initiatives, but many of these have failed to succeed. The reasons for the weaknesses of past productivity improvement efforts are remarkably consistent and are, in most cases, characterized by the following factors:

- Lack of top management commitment and effective support

- Lack of mid-management or union support

- No one personally responsible for productivity improvement

- Lack of credibility of the productivity program, including such assumptions by employees that it is "just another passing fancy" or a "management trick" to rationalize staff reductions

- Lack of sufficient innovation—"always the same thing"

- Too heavy dependence on macropolicies and overall policies without sufficient attention to different needs and different impacts of policies on various divisions, sectors, and production or service units

- Lack of clear objectives that focus on important goals and avoid wasteful digression from proceeding toward achieving the objectives

- Lack of a plan for early success, for instance, by selecting such easy and quick measures as eliminating bottlenecks or simplifying work flow

- Being operated apart from the overall business plan, outside the mainstream of the operation and, as such, largely ignored by top management

- Ad hoc, disjointed nature of the productivity effort, often resulting in improving productivity in one area at the expense of another, and usually too expensive and too time-consuming

- Poor organization or insufficient coordination

- Orientation toward the short term—reflected by a short-range focus in budgeting—solution of immediate problems while being counterproductive in the long run

- Overly narrow in scope, often focusing on measurement only

- Lack of clear objectives when developing productivity measures, without clarity of what they are intended to measure

- Focus on the measures, if developed, of direct labor only, ignoring the often overwhelming indirect labor element

- Not using the indicators or productivity measures in decision making

- Not taking corrective action after weaknesses have been identified

- Not telling staff what is expected of them

- Failing to learn from what others do, and not implementing the relevant lessons learned

- Insufficient attention to managerial and supervisory training

## CUSTOMER ORIENTATION AND FEEDBACK

One of the most significant changes in today's organizations is a shift toward customer orientation. The organization's ultimate purpose needs to be the service to its customers. The organization's mission statement, task planning, and description, as well as performance assessment, have to be defined from the perspective of the customer. Product or service design must meet the customer's needs and wants. The quality of products and services needs to be measured and analyzed from the customer's viewpoint at every stage of the production process. Data and records should be limited to those required to serve the customer's objectives. Superfluous requirements should be avoided because they could unnecessarily increase costs and delay schedule and deliveries.

Keeping customer requirements in the foreground often means major corporate structural reorganization. Successful companies have found, for instance, that the creation of "multifunctional" jobs dealing with various aspects of production or customer relations can create a customer-oriented climate of "placing the producers into the customer's shoes." It is important to make the customer's satisfaction the concern of every worker. Enabling the workers to be able to handle all relevant customer-related responsibilities makes it possible for the workers to look at their product or service from the customer's viewpoint. It is essential to reemphasize that the productivity and quality requirements should not be limited to the actual product or service provided to the customer but also to all related delivery, billing, and other administrative activities. For example, in hotels and motels, guest supplies such as towels, soap, and shampoo, shoe-shine cloth, sanitized glass covers, sanitized toilet seats, paper supplies and stationery, are very important to patrons, just as looking after guest laundry, dry cleaning, and ironing are.

The wide choice customers nowadays have among products as well as suppliers means that the establishment of a strong relationship between producers and consumers as well as the development of customer loyalty have become very important, and need to be nourished through high quality and possibly personalized service. The high quality of customer service must be ensured through continual communication with the customers so that the organization can develop a feel and knowledge of what the customer needs and wants. Two-way communication with customers is also needed to ensure that quality defects can be immediately corrected. Customer feedback can be institutionalized by planning and evaluation meetings with the customer, joint procurement quality committees, on-site visits, conferences, and industrywide quality surveys.

## OVERWHELMING IMPORTANCE OF THE HUMAN FACTOR AND EFFECTIVE MANAGEMENT OF EMPLOYEES

There seems to be unanimous agreement among successful modern organizations that all such strategic issues as productivity and quality improvement, creativity, customer satisfaction, and cost reduction hinge on the human resources, the competence and commitment of staff. Human factors, including management's commitment to their employees and their respect of the individual, determine the attitude of the workers. Employees who are served well by their organization and feel that their employers are concerned with their well-being, safety, health, and comfort are creative, interested in providing work of quality and serving the customers well. Plants, equipment, and other facilities work well only if operated by capable and satisfied workers.

Workers react favorably to knowledgeable and caring leadership. Managers, therefore, need to understand in reasonable detail the entire production or service process, know how the job needs to be done, and how to modify the workers' activity if it is necessary. The manager needs to act positively, find out where the fault lies, and help the workers bring the performance up to standard. It is usually demoralizing if they pick on every defect or select the lowest example of performance and criticize that. Virtually all workers try to do their best if they are treated well, encouraged, and helped when necessary. In order to ensure these managerial qualities, a number of companies actually survey the views of the subordinates of their manager.

Throughout our discussion we have emphasized that productivity and quality improvement are the responsibility of everybody. It has been observed, nevertheless, that it is management, including top and middle management as well as supervisors, who have the primary function of improving them. They must exercise leadership, stimulate organizational vitality, and ensure that the organization gets its job done. They must have the right attitude toward work, and stimulate their staff by their knowledge, experience, initiative, fairness, and helpfulness. They need to exercise a modern, cooperative, and open management style rather than traditional, outmoded, and secretive authoritarian practices.

Among the managerial factors that influence productivity, personnel management always stands out. The quality of staff should be optimized through effective recruiting of new staff as well as through positive direction, guidance, support, and appraisals of existing staff. When hiring new people, the manager or supervisor should learn about the applicant as much as possible before the interview. The interview itself should be as relaxed and open as possible. The applicant should be encouraged to describe his/her qualifications, strengths and weaknesses, experiences, accomplishments, and aspirations. Let him/her indicate why he/she wants to work for the organization. The interviewer should listen carefully but retain the initiative throughout the interview. The application form should preferably be looked at before the interview and notes should be taken right after the interview's completion. Questions should be asked about gaps and inconsistencies in any statements. The interview should be closed firmly but tactfully. If the applicant is clearly unqualified, the interview should not be prolonged. Alternative solutions may be suggested.

Performance interviews should also be preceded by the preparation of a written fact sheet about the employee. Points to be covered may include, among others, the quantity, quality, and timeliness of work, attendance, reliability, attitudes, physical appearance—as long as it affects the work or work environment, changing work habits, and any other points that may be relevant to the employee's productivity and general performance. Good work should be recognized, but consistently poor performance should never be covered up.

Attendance may become unsatisfactory. Absenteeism may appear, with increased frequency of time off work, minor illnesses, or repeated unjustified absences on Mondays or Fridays. Poor attendance may also appear in frequent lateness to work, extended lunch periods, or often leaving work early. The employee may become unreliable, sloppy in handling assignments, tend to put things off unnecessarily, or blame others for problems. He or she may become irritable, untidy, or show noticeable changes in normal behavior.

Performance problems may mean hazards to health and safety because of careless handling and maintenance of equipment that may lead to frequent accidents. The manager or supervisor should try to identify the reasons for the problems. Performance problems may be due to a lack of training, incompetence, incapacity because of physical, emotional, or other health reasons, or a lack motivation. If there is need for correction of performance, the supervisor should indicate what corrective steps should be taken. The discussion should be carried out in a sincere and positive manner. Confidentiality should be assured unless there is a need for involving others in the corrective steps because conveying the information to these others is essential to the solution of the problem. Also, the applicable laws need to be followed.

Managers and supervisors should endeavor to help individual employees resolve their problems and reach their fullest potential. They should listen to them carefully, treat them with dignity, respect their feelings, encourage them to voice their ideas, and make adequate time available for discussion.

# CHANGING ROLE OF MANAGERS AND SUPERVISORS

Management needs to plan, challenge old rules and procedures, innovate, coordinate the related activities of all departments, communicate the mission, goals, and objectives of the organization to all personnel, and provide their staff with the necessary time and resources for quality production. Managers need to enhance productivity through systematic strategic planning, as well as encouraging and monitoring efficiency, effectiveness, and quality. They need to select the best staff, coach them, and utilize their capabilities, motivate them, discuss and help resolve problems, use and ensure the use of statistical productivity and quality control methods, as well as ensure that all of the staff benefit from productivity and quality performance. Managers also need to communicate with the customers and users of their products or services.

Top managers must be committed to implanting a corporate culture of productivity and quality, as well as provide leadership and guidance for productivity and quality improvement. They need to personally introduce the productivity program and demonstrate that productivity improvement is a vital necessity and an important company objective. They should show by their own example that the productivity policy is serious. They have to review the productivity program periodically, actively support the work by providing resources, and hold the middle managers to their productivity targets. Junior managers should continually seek to find and introduce better procedures, help staff understand productivity, lead productive and high-quality operations, assign tasks, monitor operations, collect data, prepare progress reports, and make recommendations.

Obtaining management support sounds easier than it is. In the case of top management, its frequent isolation, its characteristically high mobility, and resistance to organizational change present great difficulties. As far as middle and lower managers are concerned, the warranted or unwarranted fear of the weakening of their power and status by such useful organizational forms as quality and productivity circles or committees, as well as apprehension of other major changes, may generate resistance. It has also been observed that management still often has the antiquated view that line workers are the primary cause of productivity and quality problems, and that there is, therefore, no need to introduce changes in other aspects of the organization or at higher levels of the hierarchy. As noted elsewhere, about 80 percent of quality problems tend to be attributable to the system, and not to production and line workers. Top managers need to reeducate those other managers who resist productivity or quality improvement.

Many changes are taking place that affect the role and needs of middle managers and first-line supervisors. Technological change has made many jobs easier but at the same time more complex. Office computers and computer-aided design and manufacturing change the nature of work, for example, by improving products and services, often reducing the cost of production, making the work easier, but also requiring considerable knowledge of both hardware and software. The new developments need more adaptable, flexible, as well as innovative methodology and procedures. The managers need to keep abreast of new technology, training, and methodology, discuss these at meetings, and help solve problems.

Improved educational standards have led to increased expectations. The trend toward group operations and strengthened labor–management cooperation has placed emphasis on the functions of middle managers and supervisors as leaders of work groups. They continue to provide leadership, direction, instruction, and guidance, as well as supervision and discipline, but more on a group basis than on a one-to-one relationship basis. Managerial philosophies are also changing. The trend is toward increased direct communication between senior management and the workers, and toward worker involvement in managerial decision making. While these developments tend to increase worker productivity, they also change the nature of the job of middle managers and first-line supervisors.

People want to participate and are usually willing to contribute to their organization's productivity efforts, but they must be informed of the benefits of productivity improvement to them, as well as to the organization. They need to be informed as to how they can participate. In the more cooperative atmosphere of today's organization, the supervisor's role as a leader and example-setter has become even more important. The supervisor must see to it that the management's goals and policies are harmonized with the legitimate expectations of employees. For example, the training of employees should not only be directed by the interest of the employer but also by that of the employee, for example, the employee's career advancement.

The new system also expects that the supervisor act as a coworker, cooperating with fellow workers in order to support them, train them, and provide guidance. The supervisor needs to be an innovator who not only listens to the suggestions and ideas of staff, but translates those into direct action if justified and meets organizational objectives. Advances in technology, management, and labor relations enable significant improvements in productivity, but the relevant new role and functions of middle managers and supervisors bring with them the need for new skills.

The required training and retraining have two major aspects, namely, awareness training and skill training. Awareness and understanding of the changes that occur and of the importance and potential of improved productivity are of utmost importance. Effective tools that can be used in awareness training have included films, case studies, and visits to organizations using the new systems. New skill training is also needed by middle managers and supervisors. The new skills will show them how to get where they want to go, once they understand the necessity and usefulness of the new system. The new skills fall into two main groups, namely, task-oriented skills and people-oriented skills.

The task-oriented skills involve technical skills, production planning, implementation, and controlling, management information systems, problem solving, decision making, organizing, administering, delegating, oral and written communication, conducting meetings, and brainstorming. Of great importance among task-oriented personal skills of middle managers and supervisors are the clear understanding of their responsibilities, planning and using time and energy efficiently and effectively, identifying priority tasks and their determinants, boiling down complex functions into simple elements, devoting attention to tasks in relation to their importance and urgency, delegating responsibility to subordinates for work that the subordinates can perform, and focusing on their own priority functions such as planning, leading, motivating, advising, and controlling.

The new forms of organization and participative management also require that middle managers and supervisors develop their people-oriented and group-related skills. The group-oriented skills are needed because many of their functions now have to be performed on a group basis, including such functions as team building, communication, work assignment, directing, guiding, advising, and motivating. They need, therefore, to be able to make effective presentations, conduct meetings, and build consensus.

# QUESTIONS

Q: 8-1   List a few major requirements of effective management of human resources.

Q: 8-2   Well-qualified and motivated workers require relatively high wages or salaries. Does this requirement cause high labor cost?

Q: 8-3   Whose responsibility is it to improve productivity?

Q: 8-4   Why is top-level commitment and support essential for successful productivity improvement?

Q: 8-5   How can the leadership ability of a manager be assessed?

Q: 8-6   Does quality improvement hinder productivity?

# SUGGESTED ANSWERS

A: 8-1   Major requirements of the effective management of human resources include leadership, support, fairness, flexibility, skill variety, training, recognition, making clear assignments, providing feedback on performance, caring about the quality of working life and safety, and so on.

A: 8-2   High wages for well-qualified and motivated workers do not cause high labor cost because their capabilities and higher productivity tend to more than offset the relatively high wages if their ability and capacity are well utilized.

A: 8-3   Everyone must aim for and contribute to productivity improvement. Nevertheless, the primary responsibility rests with management, who combine the various resources into the best possible overall results.

A: 8-4   Top-level commitment and support is essential for a successful productivity effort in order to provide a productivity role model and example to other managers and staff and to give credibility to the productivity program.

A: 8-5   The leadership ability of a manager can be assessed by his or her accomplishment of goals, and impact on the productivity of the unit.

A: 8-6   Quality improvement helps productivity because it eliminates wasteful reworking, reduces warranty cost, and increases customer satisfaction.

# 9

# There Are Many Opportunities for Improving Communication

## HOW TO IMPROVE COMMUNICATION WITHIN THE ORGANIZATION AND WITH SUPPLIERS, CUSTOMERS, AND CLIENTS

In order to take correct actions, one needs to know the underlying conditions, what is needed, what one wants, what others want, what would be the consequences of alternative actions, and so on. None of these questions can be answered without communication with superiors, staff, colleagues, customers or clients, and suppliers. One of the basic rules of communication is that we must make clear what we want to say. The other fundamental need is to listen carefully to what others say and consider carefully what they are interested in. This latter point is very often neglected.

These general rules also apply to communications about productivity. It has already been emphasized that the meaning of the word "productivity" is commonly misunderstood and misinterpreted and is, therefore, often ignored or even opposed. In order to help improve the productivity of an organization, management must create companywide productivity awareness and help employees recognize that productivity improvement is in their own interest, that their own contribution is essential for the improvement of the organization's productivity.

The achievement of all these objectives requires good communication throughout the organization. The lines of communication should be made as short as possible. Two-way communication must be ensured. Management must listen to the objectives and problems of their staff and, on the other hand, management must help employees learn how important productivity is to the organization as well as to them, and explain how they can contribute to the improvement of productivity and the success of their organization.

In order to be motivated toward productivity, employees need to be "in the know," that is, know where their organization stands in relation to their productivity and the productivity of others, particularly that of their competitors or counterparts. Information about the productivity activities within the organization should be communicated to all staff. Employees also want to know how productive and competitive their organization is now compared to the past. Is it improving or falling behind? It is virtually impossible to expect the employees' cooperation in the organization's efforts unless they have clear and understandable information about the goals, objectives, capabilities, successes, and problems of their organization. They also need to see how their own work contributes to the success of their organization. Workers also must be told what is going on in general that can affect their job.

Communication with colleagues is also important in an organization because the activity in one area often affects the activities in another area. For example, design affects production, marketing, and physical distribution. Communication is also necessary with suppliers because of the impact supplies have on production and vice versa. Satisfactory communication with customers and clients is essential because one needs to produce or provide what the customers or potential customers need or want.

Effective communication can take the form of information sharing, management and staff meetings, discussion groups, graphic displays of productivity (including quality) performance, productivity contests and rewards, speeches by senior executives, newsletters, and training sessions. Success stories need to be publicized so that successful methods can be duplicated. It is important that all workers understand and believe that the productivity movement is in their own interest and not a management ploy to make them work harder for the same pay.

# PREPARING EFFICIENT AND EFFECTIVE WRITTEN REPORTS AND ORAL PRESENTATIONS

Another form of communication that from time to time all of us need to undertake is preparing written reports and making oral presentations. In order to prepare effective reports and oral presentations, it is essential to clearly identify the objectives, that is, what you want the audience to know and why, then suggest what the audience could do about it. The interests, concerns, and questions of the audience should be anticipated. The presentation should be clear, simple, concise, honest, arranged in a logical sequence, and use words and expressions that the audience can understand. If several presentations are scheduled, for instance in a conference, the chair should ensure that each presentation is ended within the allotted time.

In written reports, it is helpful to use headings and subheadings for the sake of clarity. One should prepare an outline to ensure logical sequence and focus on the main elements. The presentation should be natural, preferably not read, and presented with confidence. At the beginning it is useful to indicate briefly what will be said, and to end the presentation by summing up the intended message, stating what action is necessary, if any, and by whom and when.

# CONDUCTING PRODUCTIVE MEETINGS

The communication skills required include the ability to conduct productive meetings. It is important that meetings be held only when needed for a specific purpose or when a team effort is required. One should keep in mind that there may be better ways to disseminate information. Keep the meetings as small as practical. If you are invited to a meeting, consider why you are invited, and if it is not evident, ask what you are expected to contribute. The meeting should be held at a convenient location. The agenda and background material of the meeting should be distributed to the participants in advance and read by them in order to be prepared for the discussion. The chair should briefly sum up the highlights at the opening. The meetings should start and end on time so that the participants do not waste their time waiting. If possible, it is preferable to limit meetings to not longer than one hour, so that they end before the participants lose attention.

Open the meeting with some introductory remarks to encourage constructive participation. Adhere to the agenda and keep the meeting on schedule. Unnecessary digressions should be avoided. If some of the items can be settled by only a few of the participants, their resolution should be assigned to those interested. From time to time it is useful to sum up the topic just concluded, and at the end of the meeting the chair should sum up the conclusions and, if any action is needed, assign them to the appropriate participants. Minutes should be kept of the meeting, and once they are finalized they should be circulated to the participants without unnecessary delay.

# HOW TO PREPARE AND CONDUCT SUCCESSFUL NEGOTIATIONS

Negotiations are among the most important forms of communication in which all of us have to engage from time to time. Negotiations are a form of joint decision making by the negotiating parties. We negotiate because we want something and are willing to offer something in exchange, for example, when we buy or sell something that does not have a fixed value or quality, when we want to determine the benefits or cost of something, and when we want to resolve differences, problems, or conflicts. We also negotiate when we want to allocate resources, take a job, hire an employee, set wages or benefits, and discuss a performance appraisal report. It is always to our benefit to negotiate effectively and efficiently. The objective of most negotiations is to determine the value, benefits, or costs of something, agree on important quality requirements or other significant details, reach agreement on taking some action, or establish a relationship between individuals or groups.

The productivity of a negotiation is determined by the value of achieving the goal as compared to the cost and effort of reaching the agreement. The first means how effective our negotiated agreement is while the second shows how efficiently we negotiated. The quality of the negotiation is best reflected by whether it provides the best deal for both parties—whether the relationship between the parties that has been developed during the negotiation is mutually satisfactory.

Effective negotiations have no standard rules. Negotiators must, however, know why they negotiate, what they and the "other side" want to achieve through the negotiation, what their main issues, interests, and limitations are, what costs and other limits will be acceptable, and whether there are any sensitive areas. It is useful to list the objectives in writing in priority of importance to the negotiator. It is important to develop a long-term plan that will ensure that the maximum of the negotiators' objectives is achieved and that their costs are minimized.

It is important to prepare negotiations carefully. Any background material and relevant previous negotiations must be thoroughly studied, and a decision must be reached on what should be included in the negotiations and what should be excluded. All expected costs need to be considered, including such visible costs as physical and material resources, as well as the human services and the necessary support services and warranties required in the future. The potential risks and responsibilities that may arise because of the negotiations should be kept in mind.

It is important for the team leader to reach a common position for the team and obtain the prior approval of superiors. The time, location, and length of the negotiations should be decided to create conditions that would benefit the negotiating team. It has to be decided who will negotiate, where, and under what circumstances. Among very important strategic considerations is the time of the negotiation, the time that is most favourable for the negotiating party, how long the negotiations should last, and by what time they should be concluded. It is usually important to have alternative plans prepared in advance, including possible trade-offs that can be offered for concessions by the other party.

It is desirable to start with the more simple and acceptable points of discussion in order to build up trust. Unnecessary digressions from the main problems should be minimized. In order to ensure common understanding of the negotiation, the proceedings should be summed up from time to time by the leader of the negotiation. The negotiators must avoid making any hasty decisions and it is, therefore, usually preferable to stick to the original plan. The closing is important and should be on a positive note.

As in all productivity matters, people are the key in negotiations. Effective negotiators need training and experience in order to be successful. They should develop and practice their negotiating skills to be effective. Negotiators benefit from having power in the negotiations. They gain power from knowing the topic and objectives thoroughly and from having a position or former position that may convey authority and credibility.

Good negotiators tend to have several typical characteristics. These include the following:

- Negotiators should always be aware that first impressions tend to be very important in negotiations.

- Negotiators must be thorough and specific, avoid vague generalities, and explain and justify what they are saying.

- In order to be credible, they need to be accurate, as well as realistic, and avoid giving the impression of exaggerating facts.

- They must be, and appear to be, honest in order to be accepted by the opponents as trustworthy.

- In order to strengthen the trust that will be needed throughout the relationship, the negotiators must be fair and not exploitative.

- They must respect the values, beliefs, and sensitivities of their opponents, and allow the opponents to save face, if necessary. Therefore one needs to explore and understand the goals, beliefs, interests, and expectations of the opponents. If some point is not clear, the negotiator should ask questions and listen carefully to what the opponents say.

- The other party's preferences should be anticipated, and how the negotiator's objective could benefit the other party should be assessed. The purpose is to end in a "win–win" situation, namely to achieve the negotiator's objectives with the other party also coming out as a "winner."

- Negotiators benefit from a convincing personality if they project self-confidence, speak clearly, not too fast, and with a strong voice.

- It is important that negotiators do not appear hostile.

The draft agreement must be thoroughly reviewed in order to ensure that it achieves its objectives, that it is clear, workable, as well as legal, and determine what it costs. It needs to set out what each party is responsible for and what the consequences and penalties are if the terms of the agreement are not implemented. Before the agreement is finalized, the team leader needs to convince her superiors of the benefits and reasonableness of the tentative agreement. A good agreement must be flexible enough to permit later adjustments if desired.

After the final agreement is reached, the results should be communicated to all interested parties. It is advantageous to emphasize in the announcement the positive aspects of the agreement. Attacking the sensibilities of the other negotiating party should be avoided. To be effective, the agreement must be implemented through a continuing cooperative relationship with the other party. As with any other effective and efficient action, the performance of the negotiated agreement must be monitored, measured, evaluated, and the findings fed back to the negotiators, as well as their superiors and relevant interested parties, such as the rank and file.

# QUESTIONS

Q: 9-1   Suggest examples of how communication and feedback of information affect productivity.

Q: 9-2   How can communications be made more productive?

Q: 9-3   What needs to be communicated to employees in order to ensure their cooperation?

Q: 9-4   List some of the main requirements of writing effective reports.

Q: 9-5    What are the main requirements of a good oral presentation?

Q: 9-6    Mention some requirements of leading a good meeting.

Q: 9-7    List some of the main characteristics of effective negotiators.

# SUGGESTED ANSWERS

A: 9-1    Proper communication and feedback of information can improve productivity and quality because it helps prevent wrong decisions, avoids duplication in searching for the same information, and helps to take all necessary information into consideration. Feedback on productivity (including quality) problems enables correction of the problems, and feedback on performance provides effective guidance to employees.

A: 9-2    The productivity of communications can be enhanced by ensuring that the necessary communications do take place, that unnecessary communications are eliminated, and that too-complex communications are simplified.

A: 9-3    Employees will cooperate if they are given clear, consistent, and frequent communication about the organization's goals, objectives, productivity plans, and achievements. Employees want to know what they are contributing and why, and how they are doing. Employees will also be more productive if you seek their suggestions because this will indicate to them that their work is important for the organization's success.

A: 9-4    Effective reports require a clear definition of the report's objective, that is, what one wants the audience to know and why. The interest of the audience must be taken into consideration. The write-up should be clear, concise, simple, relevant, well balanced, well organized, and not too long. A brief summary in the conclusion is usually helpful.

A: 9-5    In a good oral presentation the objective should be set out at the beginning, and it is useful to start with an attention-getter. The presentation should be clear, simple, concise, and not too long. It needs to be organized into a logical sequence. At the end it is usually effective to sum up what was said and what the conclusions are.

A: 9-6    In order to conduct an effective meeting, you should have a clear understanding of the issue, ensure participation, stay on the subject, identify decisions, determine clear roles in action plans, and take minutes of assignments.

A: 9-7    Effective negotiators must be forceful, thorough, credible, realistic, specific, honest, fair, and not appear hostile.

# 10

# Effective Methods of Motivating for Results

## MOTIVATION IS ESSENTIAL FOR ENHANCING PRODUCTIVITY AND QUALITY

The success of productivity programs depends to a large extent on the motivation of all staff members. Experience shows that equitable pay and proper incentives lead to higher productivity because they attract and retain good workers and stimulate them to do an extra good job. They reduce absenteeism and turnover of desirable employees that would, in turn, result in lower productivity and quality and cause increased costs. People perform better if the rewards for their work are tied to their actual performance. Incentives, therefore, are often used to stimulate them. It is important to design reward systems so that they adequately correlate productivity and compensation.

In order to be able to motivate their staff, managers and supervisors should make a genuine effort to understand their staff, know their strengths and weaknesses, their goals, the primary factors of their motivation, as well as the causes of their frustrations and dissatisfactions. For these reasons, employers should encourage their staff to ask questions and should answer them clearly and honestly.

The effectiveness of reward systems depends on offering rewards that the employees value. The system must be based on simple, credible, fair, and repeatable evaluation methods. In order to be satisfied, workers need to know what they are to do and how they are doing. It should always be kept in mind that no individual can be motivated to do work for which he or she does not have the necessary skills. They need, therefore, a clear job description in which their tasks are identifiable. Most workers, particularly professionals, like autonomy, that is, to be told what is expected of them and then be left alone to do it the best way they can. Most workers like variety in their work, for which they can use a variety of skills.

When considering incentives, it should be determined whether the incentives would achieve their goals to enhance performance and what their costs would be compared to their benefits. It should also be clarified whether the planned incentives would have negative consequences. Are there any better incentive solutions? It also needs to be decided whether individual or group incentives would best achieve the objectives.

# NONFINANCIAL INCENTIVES ARE VERY VALUABLE MOTIVATORS

Incentives and rewards can be nonfinancial or financial. One must, however, provide incentives that the employees value. Nonfinancial rewards, awards, or other incentives are often valued by employees just as highly as financial rewards. Most workers work better in jobs they are interested in and that challenge them. Most like to see the contribution they make to their organization and its products or services. They appreciate being informed by their supervisors on the progress of their organization, their role in it, and their importance to the organization.

For these reasons, they like a job designed to represent a clear function, a job the product of which can be identified and is significant. Creative people particularly appreciate highly challenging job assignments with variety in their work, with a minimum of boring, trivial, or repetitive tasks. Most tend to like various forms of job enrichment such as job rotation to widen operational knowledge in order to understand the organization's various functions and programs. For this reason many employees favor learning different functions and skills within and outside their own program. Key people tend to like special assignments in addition to their normal work. Most employees are looking forward to opportunities for advancement.

It is important not to take excellent work for granted. People are stimulated to do their best if they are shown that their supervisors are interested in their success, if they are given feedback on how they perform, and if their achievements are recognized. Individuals are motivated if they feel that they are doing well. It is important, therefore, that they can establish objectives for themselves that they can reach. They welcome effective and fair measures of performance, equitable promotion policies that reward higher productivity, and special career development opportunities.

High performance and high quality of work should be formally recognized and if possible rewarded, perhaps by giving some awards. Most employees are encouraged by formal recognition, bonuses, quality of work life improvements, trips, or other awards. The workers should be thanked for their work, possibly in public such as at meetings, in newsletters, letters by the boss, or phone calls to the employees, and so on. The communication not only enhances the employee's pride but also tends to provide employees with opportunities to discuss and further advance productivity and quality improvement. Managers and supervisors should recommend superior employees for special awards and make award presentations for outstanding performance. Both individuals and teams can be rewarded. Attention must be paid to ensure that the rewards are fair and for real contributions.

You can generate many useful productivity improvement ideas if you give your employees an opportunity to express themselves and make suggestions to improve your operation. Employees appreciate when their employer listens to their views, interests, and problems, and follows up on employee suggestions in a timely fashion. It has been observed that in such circumstances the number of suggestions tends to rise over time and that the adoption rate also increases.

Worthwhile ideas should be rewarded at least with small awards. If employees' ideas can not be implemented, they should still be thanked for them and it should be explained why they can not be used, at least for now.

Improved quality of work life is greatly appreciated and can contribute to higher productivity. Such factors include congenial working conditions, acceptance by peers, reasonable hours, better light and ventilation, rooms with windows, comfort, temperature control, industrial hygiene, and suppression of excessive noise. These also enhance the vitality of the organization. Security and safety at the workplace are important aspects of the quality of work life that enhance productivity and quality in the twenty-first century. Protection from fire is essential, particularly in high-rise office buildings and crowded areas. The provision of adequate fire extinguishers and safe evacuation routes is essential. All staff need to be informed and trained in safety procedures.

The workplace needs to be protected by adequate physical barriers and, where appropriate, people entering the premises should be screened for security. In high-security areas, identification checks, passes, and electronic monitoring may be required. Employees need to be trained to be alert for unknown or unauthorized persons in these areas and to notify security personnel of such observations. Personnel, belongings, as well as official items should be kept safe.

One of the effective rewards for promising, highly productive employees is the provision of training and retraining opportunities in view of changing technologies and managerial methods, shifting job opportunities, and with regard to the interests of the workers. For these purposes, employees like to attend development seminars or conferences. Most employees appreciate advanced managerial training if they have not received it earlier. In many lines of work, it is effective to provide equitable parallel managerial and technical promotion ladders. If there is an intention to authorize training opportunities, it is useful to ask the employees what training they are interested in and, if the request is reasonable, authorizing it.

Obviously, not all programs suit all organizations. In the public sector nonfinancial incentives are of particular importance because the opportunities for financial incentives are rather limited.

## INVOLVEMENT AND PARTICIPATION ARE ESSENTIAL NONFINANCIAL MOTIVATORS

Nonfinancial incentives include formal recognition, enhanced titles, greater involvement in organizational activities, and being kept "in the know" by being regularly provided with information about the performance of the organization and relevant business information. Employees value being greatly involved in the productivity and other programs that affect them, as well as having their suggestions and ideas seriously followed up. Additionally, having the security of their jobs assured by their fine job performance is also valued.

People today want to participate and contribute to the company, and the contribution of individuals to productivity improvement can be very significant because very often, if not most of the time, it is the employee himself or herself who can best judge how his or her work can be improved. Employees, therefore, must be involved in the development of productivity improvement plans. If the employees are involved in developing the plans, they also have a vested interest in trying to ensure their success.

Such involvement enables the organization to use the employees' skills, knowledge, and abilities to do their work most effectively. Enabling employees to participate in all matters that affect them, including the advancement of the productivity of their organization, gives the employees a feeling of belonging and a common purpose with their organization. Their full participation brings about higher overall productivity and improved quality of the goods and/or services they provide.

Worker involvement can take many forms, for example, suggestion systems or various types of teamwork. The latter can be brought about in several ways, such as through productivity problem-solving teams or quality improvement committees at the corporate and/or operating levels. In many cases, small-group activities such as quality circles or discussion or suggestion teams are practical means of bringing about the required communication, participation, and cooperation.

In other cases labor–management committees—distinct and separate from the collective bargaining groups—have been found to ensure participation by everyone. Employee involvement in increasing productivity and quality necessitates the positive participation of employee unions and other employee organizations. These activities are productivity-related, in addition to—and not replacing—union responsibilities in collective bargaining. At the same time, it is necessary that leaders of organized workers accept a greater responsibility for the successful operation of the organization in which their members work. This enhances their members' long-term interests, including job security and a share of organizational success.

Worker involvement has become generally recognized as an essential ingredient of productivity and quality improvement. In the traditional approach, suited to large-volume production of basic goods or services, standardized production methods were used. Labor cost was minimized but there was very little scope for individual creativity or other individual contributions to higher productivity, quality, or better service to the customer. Today, in many modern and successful organizations, each worker can get involved in ascertaining customer requirements, setting objectives, determining and keeping under examination the best ways of the individual work process, identifying problems, suggesting solutions, or actually resolving those problems and making improvements within the general policy framework.

In order to meet the growing need for more-customized, small-batch production of high-quality goods and services, the involvement, initiative, interest, creativity, and care of the workers are paramount. If the workers are involved in the process from planning through production to service to the customer, workers are much more likely to be committed to the success of the objectives of the entire organization. Workers accept and support change much more fully and voluntarily if the change includes them from the start.

Organizations have recognized that no single person can do all the necessary work. Teamwork helps cooperation across functions and increases communication as well as understanding between the employees involved. Participants of the team have common objectives and need to be committed to their accomplishment. The members of the team need all the information that may influence their work, as complete and accurate as possible.

Teamwork, therefore, has become more and more an important necessity. One form of teamwork is quality teams. While these can take many forms, they tend to take the shape of small teams of, say, 10 to 15 employees each. Teams are typically formed on a voluntary basis, but successful large companies have reported that the majority of their workers choose to participate. Team meetings are often held for one hour per week. Successful teams tend to apply systematic problem-solving techniques based on heavy use of data collection and analysis. When quality problems are identified, subcommittees may be appointed to develop corrective action through periodic meetings. It may take days, several weeks, or months to find the solution. Once it is found, it may then be taught to the entire workforce.

The team approach, based on statistical quality control, was made popular in the 1950s through quality circles. Each of these circles involved 10 to 15 shop floor workers who cooperated in trying to find ways to improve the quality of their work. Contests were eventually organized among quality circles within an organization and among various enterprises, culminating in national contests. The winners were given significant rewards, trips, and so on. The idea of quality circles has been adopted in many countries. The concept of quality circles was a simple, standard idea that could be implemented by management relatively easily and proved to be successful in improving quality at the lowest levels of the organization. But it did not require change within management.

The establishment of quality circles has not remedied higher-level and overall corporate quality problems, such as quality management and interdepartmental interface and cooperation, because top and middle management as well as corporate support staff were not included in the quality circles movement. The realization of this serious shortcoming has led to organizationwide productivity and quality improvement goals and programs. Many leading successful organizations are now operating such programs, run under various names. In these programs, all employees including management and professional staff are involved. Therefore, organizationwide involvement in productivity and quality improvement is both a very powerful incentive and method for productivity and quality improvement.

# FINANCIAL INCENTIVES, PROFIT SHARING, ESOP

In addition to individual financial incentives, there are also group incentives that exist in many forms. Group incentives have the advantage of promoting cooperation among employees.

*Profit sharing* is one type of group financial incentive method where a part of the company's profit is distributed to all or some employees in the form of a bonus, usually once or twice a year, in addition to regular pay. Several sources, including Edward M. Coates III of the U.S. Bureau of Labor Statistics, credit Albert Gallatin, a Swiss-born American, as the first to introduce a profit sharing plan, in the 1790s, into his glassworks in New Geneva, Pennsylvania.[1] Others followed, and since then profit sharing has spread around the world. In their recent survey, the Profit Sharing/401k Council of America reported the existence of 1106 plans with more than six million participants and $500 billion in plan assets in the 2005 plan year.[2]

Profit sharing is a recognition and reward for the efforts and contribution of employees to the success of the company. A great variety of profit sharing plans exist. Specific designs depend on the objective and culture of the company, and whether the plan is designed, for example, to increase profitability, productivity, or employee loyalty. Other plan objectives could be to improve labor–management relations, reduce employee turnover and conflicts, foster team building, provide retirement income, better job security, or to help attract top recruits. The plans can be cash plans, deferred plans, or a combination of both.

In cash plans, bonuses are paid to employees in cash or company shares once the company's profit is known, provided that there is a profit. In deferred plans, the bonuses are paid into a trust fund and kept there until a later point in time, such as when the employee leaves the company or retires. In combination plans part of the bonus is paid out immediately and the rest is put into the trust fund for a later payout.

In the United States one often hears of 401(k) plans. These are very popular "employer sponsored defined contribution plans" that are deferred plans, with payout deferred typically till retirement; however, these plans may or may not have a profit-sharing component.[3] For example, a 401(k) plan may have employee and employer contributions, but the employer contributions may not be based on profit but rather may simply match the amount of funds contributed by the employee. In that case, the plan is not a profit-sharing plan. Conversely, a 401(k) could have a profit-sharing component—it depends on the specifics of the particular 401(k) plan.[3]

Some profit-sharing plans cover all or most employees in order to enhance teamwork within the company. The exclusions may be such employees as salespersons who may already be rewarded for their revenue-generating contributions. Other plans may only be limited to senior managers. Most plans limit eligibility for participation in the profit sharing plan until after a short waiting period, ranging from a few months to several years, in order to provide an incentive for employees to stay with the company.

The contribution of the company must coordinate the company's objectives and ability to pay with what the employees tend to consider satisfactory. Some assume that employees expect at least five percent of their pay as a bonus. Ideal "target bonuses" have been considered by some to be between five to 10 percent of an employee's pay, in good economic times. The Profit

Sharing Council of America found that in 2005 the company contribution in profit sharing plans averaged 9.4 percent of pay, while the company contribution in 401(k) plans averaged 2.8 percent of pay.[4] Nonetheless, companies don't want to reduce profits below a certain amount per share of common stock outstanding. In some companies the percentage of operating earnings (that is, operating revenues minus operating expenses) paid to the bonus pool varies depending on the ratio of operating margin, which is the operating earnings divided by the operating revenues.

The profit-sharing plan needs to define how the company's contributions are allocated among individual employees. The most popular allocation is based on employee earnings, but other plans also take years of service and possible employee contributions into consideration.

Deferred profit-sharing plans may prescribe full and immediate vesting or gradual vesting. The vesting rules determine when and how much each employee can claim from his or her trust account. In full and immediate vesting the amount in the fund belongs to the employee (usually after a short waiting period). In gradual vesting the number of years of work at the company generally determines how much an employee can claim. The payments may be in a lump sum or periodic payments. Companies planning to establish profit-sharing plans need to consider the tax laws, which may provide tax breaks or obligations. The administration of the plans may be done internally or with the help of outside experts.

Management can improve the productivity–profit sharing reward connection throughout the year by productivity improvement efforts, productivity teams, newsletters, meetings, and training that explains how to improve productivity and how productivity affects profitability.

*Employee stock ownership plans* (ESOPs) are another group of financial incentives that encourage qualifying employees to buy shares of the company in which they work. In some countries, the government assists the creation of ESOPs through tax laws to promote employee share ownership. You may wish to inquire about the current rules in your area. Just as there is a great variety of profit sharing plans, there are also many kinds of ESOPs, including stock purchase plans, stock options plans, and so on. The employees may or may not contribute to the purchase of ESOP shares, but it has an important psychological impact if the employees spend some of their earnings to buy their company's shares within the ESOP. These plans may encourage productivity improvement because the workers become part-owners of the company and have, therefore, an interest in the company's success or difficulties. The employees' productivity improvement efforts can be enhanced if combined with a participative management program, training, or another form of employee involvement.

# PRODUCTIVITY GAINSHARING

In productivity incentive plans the bonuses are tied to measured productivity improvement. The incentives can be individual or group incentives. The most frequently used individual financial incentive system is the system of payment by result, such as piece rates. However, individual incentives can be unfavourable by being disruptive to teamwork and team spirit.

For this reason, many successful productivity incentive plans have been designed for groups of workers. A popular group of incentive plans are the so-called productivity gainsharing plans, often simply called gainsharing plans. The improved use of labor and other inputs enhances productivity and profits and results in financial gains. The gainsharing plans recognize that the gains are the results of the work of labor and management and, therefore, share the gains between the company and its employees. Typically, all employees are paid a group bonus instead of individual bonuses, which strengthens the team spirit and helps align company goals and employee objectives. The plans make it clear that when the workers involved achieve certain improvements in the company's productivity, they receive bonuses. The plan's specific objectives may include improved productivity such as greater output per input, the reduction of the defect rate, the reduction of cost of materials and energy, as well as enhancing the safety of employees, including the reduction of injuries.

The objectives and principles of gainsharing plans must be clearly understood and supported by the workers. Good communication between management and labor is very important for the success of gainsharing. It is important that the employees be involved in the design of the plan, its implementation, and periodic evaluation and revision. Visible top management support is essential. Unions also should be involved in gainsharing negotiations. However, the union representatives on the gainsharing committees should preferably not be the same as those on collective bargaining teams, because the gainsharing bonuses are the result of productivity improvement while collective bargaining is different and deals with wages, salaries, and other matters covered by the collective agreement.

It is also very important that the participants in a gainsharing plan understand from the very beginning of designing the plan that they can not expect productivity bonuses every period so that they don't become disappointed. Effective gainsharing plans are clear, fair, simple, and easy to understand. When the plan is initially being designed, it is desirable to include a process for any future modifications. Designing a good gainsharing plan involves a lot of complex considerations and, therefore, it is advisable to use the help of experienced experts.

Successful gainsharing plans typically improve labor relations and communication between management and labor, bring about favorable changes in managerial behavior, boost employee morale, reduce turnover, absenteeism, and grievances, and induce employees to cooperate for the purpose of improving the group's productivity. The industries most suited to gainsharing plans are those where labor input represents a significant part of costs, where output is clearly quantifiable and measurable, and whose composition of output is fairly stable.

The gainsharing "bonuses" are for productivity improvement and paid on top of regular wage or salary payments. In gainsharing plans the participants receive a bonus if they achieve a measurable productivity improvement from a base period according to a predetermined formula. Having a preset formula means that employees know how they will be rewarded and what they can expect. It is important that the gainsharing bonuses are calculated and paid soon after the period measured is over, for example monthly. It should also be determined in the design of the plan how to handle different performance in different product lines, as well as the calculation of costs in companies where some other costs significantly exceed labor costs.

The formula, which is closely linked to productivity performance, is usually based on financial rather then physical measures. In order to eliminate inflationary distortions and clearly reflect productivity changes, all currency values need to be expressed in base period values. This is usually done by using base period unit prices or by using current values and dividing them by the relevant price index using a base period price index of 100. Nevertheless, some plans ignore the distortions in order to make the calculations simple. The formula must be clear and easily understood, adjustable as conditions change, and easy to administer. Therefore, in the typical gainsharing plans the base year (or month) productivity is first calculated, and the improvement from the base figure determines the productivity gain.

For example, if the person-hours—or their wage equivalents—calculated in accordance with the base (period) productivity ratio are more than the actual hours required, there is a "productivity savings" that can be shared between the company and its employees. That is, if the current output is produced with less labor and possibly fewer other resources than in the base period, the gains achieved by the savings are shared between labor and management according to the specified formula. Improvements in quality, safety, or other company objectives can also be included as "gains" because they also influence the output/input relationship, for example, fewer accidents mean less loss, less downtime, and therefore input cost savings.

In most gainsharing plans, the workers are enabled and encouraged to get involved and make productivity improvement suggestions, often through formalized suggestion systems or productivity team meetings.

Structured employee involvement is often considered preferable, although plans can be designed where employee involvement is left more open to work team choices.

Various types of productivity gainsharing plans are in use. They differ mainly in the methods and formulas of calculating how productivity improvement is measured, how the benefits of productivity gains (or losses) are shared, and the type and level of employee involvement. There are various ways in which the productivity bonuses can be allocated among employees. Bonuses may be shared on the basis of earnings, years of service, or a combination of several factors.

Gainsharing plans are typically based on a formula that compares current productivity with that of a base period. The base period can be permanent or adjusted from time to time. It is advisable to select a base period that was a fairly typical year in which profits were reported. For the purpose of easier understanding, the inverse of the usual productivity ratio is used, namely, input over output, where input means hours of labor, cost of labor, or the latter combined with other inputs such as materials or energy used. Output means the number of units produced, number of orders, gross output, or value added.

A basic and simple plan is the Scanlon plan. It was introduced in the 1930s by Joseph Scanlon as a plan to help save the struggling steel company that employed him.[5] The Scanlon plan is based on payroll cost per net sales value of production. There are also other plans, including custom-made or hybrid plans.

The gainsharing formula of the Scanlon plan is, therefore, financial. The actual percentages of how much of the gain goes to the company and how much to the workers are determined by the company and employees when the plan is designed. In Scanlon-type plans 50 to 75 percent tends to go to the workers. In most plans, the allocation of the bonus pool among individual employees is usually made on the basis of relative wages or salaries.

Often in gainsharing plans, part of the employees' bonus share, say 20 to 25 percent, is put into a reserve fund in order to even out the effects of productivity fluctuations. At the end of a longer period, say a year, the leftover in the reserve fund is typically distributed to the employees. The plan, therefore, needs to clearly spell out how and when bonuses and reserves will be distributed.

The Scanlon plan is probably best known for emphasizing involvement and identity of employees with the company, thereby fostering productivity improvement. The typical Scanlon plan provides for a formal employee involvement system that solicits, considers, and implements employee suggestions. If the implementation exceeds a certain cost limit, the suggestion is referred to a senior committee to consider the benefits and costs of the suggestion, and whether or not it should be implemented. The Scanlon suggestion system differs from a conventional suggestion system in that individual suggestions are not rewarded individually; rather the entire group shares the (Scanlon) bonus reward.

An example of typical Scanlon plan calculations is shown in Figure 10.1. The computation starts with the calculation of the Scanlon base ratio, which is the total payroll cost divided by the total value of production, generally in the most recent three to five years of the company. Periods of major technical change need to be avoided in calculating the base ratio. Bonuses are paid whenever productivity is higher than it was in the base period, that is, payroll costs are a smaller percentage of the total value of production than in the base period. Of the total bonus pool, an agreed percentage goes to the company. Another goes to the reserve fund, and a percentage is paid to each employee.[5]

The example shown in Figure 10.2 illustrates the Scanlon bonus report. It is assumed that in company X the plan allocates 60 percent of the bonus to employees and 40 percent to the company, and 20 percent of the employees' bonus goes into the reserve fund. In Figure 10.2, $11,520 (19.2 percent of the total bonus pool for immediate distribution) is distributed to the employees.

| (Base period in dollars) | |
| --- | --- |
| Sales | 200,000 |
| Less returned sales | 10,000 |
| Net sales | 190,000 |
| Inventory change (increase) | 10,000 |
| Net sales value of production | 200,000 |
| Labor | |
| Salaries (indirect labor) | 20,000 |
| Wages (direct labor) | 20,000 |
| Vacation pay | 8,000 |
| Other fringe benefits | 12,000 |
| Total labor costs | 60,000 |
| Other expenses | 50,000 |
| Supplies | 40,000 |
| Miscellaneous | 25,000 |
| Total other expenses | 115,000 |
| Profit | 25,000 |

$$\text{Scanlon base ratio} = \frac{\text{Total payroll costs}}{\text{Net sales value of production}} = \frac{\$60,000}{\$200,000} = .30$$

**Figure 10.1**   Scanlon bonus computation for company X.

| | |
|---|---:|
| 1. Scanlon base ratio............................................................................. | .30 |
| 2. Actual sales value of production for January..................................... | 280,000 |
| 3. Expected payroll costs (1. × 2.) or (.30 × 280,000) ......................... | 84,000 |
| 4. Actual payroll costs ......................................................................... | 60,000 |
| 5. Bonus pool (expected – actual costs) = (84,000 – 60,000)............. | 24,000 |
| 6. Share of bonus given to company = 40% × 24,000 ......................... | 9,600 |
| 7. Employees' bonus share = 60% × 24,000....................................... | 14,400 |
| 8. Reserve fund = 20% × 14,400 = (20% × employees' bonus) ......... | 2,880 |
| 9. Bonus for immediate distribution = 14,400 – 2,880........................ | 11,520 |
| 10. Bonus for each employee as a percentage of pay for<br>    January = (11,520 bonus/60,000 actual payroll costs).................. | 19.2% |

**Figure 10.2**   Company X bonus report for January (current year).

Therefore, a specific employee who earns $3000 in wages per month will gain a Scanlon bonus of 19.2 percent, that is $3000 × 19.2% = $576. The bonus should be issued in a separate "Scanlon" or "productivity improvement" bonus check and not included in the regular paycheck. Year-end surpluses in the reserve fund are usually distributed to the employees.

Similarly, losses are shared according to the same split as bonuses, and the employees' share of losses is charged to the reserve fund. The method of handling possible losses exceeding the reserve fund and the year-end distribution should be determined when the plan is designed or revised, though often the company absorbs it.

Also, if your company is unionized, preferably the collective agreement and bargaining process should be kept separate from the Scanlon plan and done by different people.

In concluding this chapter, it must be kept in mind that productivity gainsharing plans are just one of the many types of programs and tools designed to motivate employees and to raise productivity in an organization. While such plans may not be suitable for all organizations, they deserve consideration because when properly designed, implemented, and administered, they can effectively contribute to improved organizational productivity and greater employee satisfaction.

# QUESTIONS

Q: 10-1  How can you motivate employees?

Q: 10-2  Suggest ways managers can motivate their staff.

Q: 10-3  List some nonfinancial incentives to stimulate productivity and quality.

Q: 10-4  Why should productivity suggestion schemes be part of a productivity gainsharing system?

Q: 10-5  When designing gainsharing plans, should the employees involved participate?

Q: 10-6  What are some critical factors of successful productivity gainsharing plans?

# SUGGESTED ANSWERS

A: 10-1  You can motivate employees with rewards and bonuses and other motivating factors such as giving employees an opportunity to use their skills, involving employees in productivity improvement efforts and decisions that affect them, praise for a job or task well done, giving suitable latitude to employees in accomplishing their tasks, training for better jobs, career guidance, and opportunities for advancement.

A: 10-2  Managers can motivate their staff by:

- Using the motivating factors listed above in A: 10-1

- Being a role model

- Showing or instructing their staff in what needs to be done

- Being helpful when needed

- Communicating that employees are valued and that their work is important and appreciated

A: 10-3  Some of the nonfinancial incentives to stimulate productivity and quality are:

- Employee involvement

- Improving the quality of working life

- Recognizing achievements

- Appreciating good work

- Providing nonfinancial incentives and rewards that the employees value

A: 10-4 Productivity suggestion schemes should be part of productivity gainsharing systems because productivity improvement is usually much greater when it is consciously attempted than when it is not. As a result, the gains to be shared are usually much more substantial if productivity is increased through suggestions. In addition, productivity suggestion systems keep the productivity improvement objective always alive and in the minds of the employees.

A: 10-5 Yes, while the methods of the various gainsharing plans vary, employee involvement right from the beginning of the gainsharing plan is very important and has the advantage that the workers have a sense of "ownership" of the plan and are more likely to accept its operation. Also, if they understand that the objective of the plan is to increase productivity, quality, and company competitiveness, as well as to reward achievement of these objectives, they are more likely to endeavor to contribute to the productivity and quality objectives.

A: 10-6 Some critical factors of successful productivity gainsharing plans are:

- Common interest of management and staff

- Fair sharing of productivity gains as agreed upon by management and staff

- Good communication between management and labor (and staff)

- Mutual respect between management and staff

- Involvement of employees in developing and revising the gainsharing plan

- Payment of bonuses as closely as possible to the time of the productivity improvement, such as monthly, and in a separate "productivity improvement" check or payment

- Enabling and encouraging employees to make productivity improvement suggestions

# 11

# The Need for Continual Training and Retraining

## TRAINING IS ESSENTIAL FOR SUCCESS IN THE TWENTY-FIRST CENTURY

Training combines instruction and practice for the purpose of providing trainees with the knowledge, skills, and experience needed to perform their jobs effectively and efficiently. One can do a job only if one knows how to do it. The training activities should ensure the development of the individuals in your organization at the same time as they enhance your organization's performance. As technology and jobs keep changing continually, the need for "continual education" or retraining becomes evident. Being updated on new processes and technology is necessary to ensure productive and high-quality operations. The necessity of looking forward also creates need for development that is based on future organizational needs for specific knowledge, skills, or experience. Development is planned growth in the qualifications of employees so that they may assume more complex duties and responsibilities at some time in the future.

When you want to decide whether training is needed, you must identify what performance problems you have in view of your objectives. Then it must be decided whether it is training that is necessary and, if so, what kind of training program is appropriate. The problem at hand may not be a training problem at all. If the employees do not know what to do, training is required. If, however, the employees are not motivated to do it right, training them will not help. The supervisors must determine why the employees are not motivated and resolve that problem. If the performance problem is caused by people being prevented from doing their job by not providing them with the necessary equipment, facilities, or time, it is not training that is needed but the provision of the right tools.

The need for training and development of the workforce is universal. Even the smaller company can not afford to neglect the training and development of its workforce if it is to remain or become competitive. The training activities must be planned carefully and tailored to the various types of staff to ensure that the training contributes to the better performance and success of the organization.

The organization has to set up its training objectives, and its employees should be familiar with them. These should include the company's productivity and quality objectives and standards. The standards must ensure that the customers' requirements are met. Employees should be trained to meet these preset standards. The quality training should include methods of how to reduce defects, errors, and waste. Positive and negative feedback from customers should be communicated without delay to workers and their supervisors.

The priority criteria for defining training needs include:

- Is there any evidence of a performance problem?

- Is it a basic need?

- Is the need organizationwide?

- Is the requirement urgent?

- Is training the right strategy to solve the performance problem?

- How should one choose the appropriate training process?

- Are the training needs mainly human–managerial–organizational or are they essentially physical–capital–technological?

- How quickly can results be expected?

- How do the costs of the training compare with its expected benefits?

- Are there ready-made programs available for the necessary training?

- How will you evaluate whether the training strategy works?

To teach improvement in both areas—the essentially human side and the mainly technological one—is important. In our experience, company after company in country after country has found that it takes much more than technology to be, become, or remain competitive in today's markets. A well-trained, flexible, interested, well-motivated, and committed workforce is necessary to be able to increase productivity, quality, and competitiveness.

As it is a precondition to provide high-quality goods and services, the selection of proper employees is essential. Supervisors and managers need to prepare for job interviews carefully, read and evaluate the application before the interview, and assess the candidate's qualifications and experience related to the job. During the interview the interviewer should avoid trying to sell the job, but rather should ask questions and concentrate on assessing the applicant's suitability for the job.

Managers who interview candidates should be trained to:

- Define the skills and attitudes required to do the job

- Ask appropriate questions based on the job requirements, application, and resume

- Ask the same general questions of all candidates

- Ask the applicant about his or her relevant accomplishments and experience

- Ask what the applicant knows about the job applied for and about the organization

- Ask them why they are interested in the job

Training is required to meet the individual needs of the new employee, when changes occur in job content, when performance deficiencies are observed, and when occupational upgrading is required. Emphasis also needs to be placed on the training of junior staff so that they can take over many functions from more senior personnel, for example semiprofessional staff who take over suitable functions from professionals, such as nurses and paramedics from doctors. Future supervisors need to be given the necessary training.

When training needs are indicated by performance problems, it is necessary to determine whether the employees could do the job properly if they had to, whether they have ever performed the task properly, and whether they know what standard of performance is expected of them. If the employee is not familiar with the work, you must provide the necessary training. If the employee is still not fully capable to do the new task but able to learn, provide further training.

In addition to his or her own work, each employee must know how he or she fits into the whole organization. This basic training may not need to take longer than a few hours or it may take several days. It should then be reinforced by on-the-job training by an experienced employee who can demonstrate or supervise practical examples or duties and answer related questions.

Individual training can be supplemented by group training and demonstration sessions as required when, for instance, new policies or machines are introduced.

To improve productivity and training, managers or supervisors need to determine what workers do when they do the job in question properly, what the problem workers are doing that they should not be doing or, respectively, what they are not doing that they should be doing. Weaknesses in performance must be clearly identified in cooperation with managers, supervisors, and the workers involved. Such problem identification should then lead to effective corrective training and development methods. What should be done? Is it being done? Also, managers and supervisors need to provide refresher courses for their staff as needed.

Using an example from the accommodations industry, employee performance and overall hotel performance depend on the knowledge of how to fulfill task and job requirements and meet milestones and objectives. Hospitality workers often can not perform their tasks properly, and tend to quit frequently unless they are formally introduced to the work and fellow workers, and are provided proper orientation and training.

Training curricula need to include the strengthening of basic skills and capabilities. One fundamental training need that has not been identified as extremely important until recently is the upgrading of the functional literacy of many employees. The problem of functional illiteracy of many workers and the widespread nature of this problem have too often been unrecognized. By functional literacy we mean that the employee is sufficiently competent in reading, writing, and basic arithmetic to be able to read and understand the instructions needed to correctly carry out the job well, follow instructions if given a new task or assigned to a new job, and to effectively participate in training and retraining courses. For quality production, functional literacy in basic statistical methods and basic computer skills is also necessary. Weaknesses of functional literacy should be watched for and recognized by supervisors and managers—workers often tend to hide such weaknesses in self-defense—and the workers should be provided with the necessary teaching and training. Employees also need to be taught the value of quality work, the concepts of productivity and cost, as well as the ways of analytical thinking, such as why a task or job has been done in a particular way and how it impacts the company, employee, and other related stakeholders.

Participant acceptance is an extremely important prerequisite of effective training. Learning is a receptive process. We won't learn what we don't want to learn or what we are not interested in learning. Therefore, the training must be of interest to the individual, as well as to the company, and the trainees need to be convinced that the training is important to them and benefits them. Employees must be encouraged to take advantage of the training opportunities. They should be trained to continually evaluate their own performance so that errors are not repeated and their performance can be continually improved.

In order to ensure successful training, appropriate training facilities need to be provided and these should be made available to all relevant staff. Training time should be reserved in the work schedule. The training of employees needs to be coordinated with the physical facilities. For instance, new equipment must be accompanied without delay by adequate training in its full use. The results of the training should be measured by indicators of improvement in productivity, quality, quantity, time, and cost.

# TRAINING IN PRODUCTIVITY IS A SPECIAL NEED

Technological, demographic, and other changes have led to a rapid increase in global competition, and all these factors have multiplied the need for improved productivity. Unfortunately, the teaching of productivity still does not form part of hardly any educational system. In order to improve productivity, managers, supervisors, and, in fact, all employees need to be trained on the job or through special seminars and courses in productivity-related skills and techniques. The importance of upgrading management skills becomes accepted if managers are made aware of the opportunities that exist for improving productivity and profits. In hotels, for example, managers need education in productive hotel operation, not only marketing, which is often considered their primary concern. The productivity objective should cover all operations.

Some of the following points have already been or will be dealt with, but in view of their overall importance these major points are discussed here in a different context. It is primarily managers and supervisors who need to learn productivity improvement techniques. These persons must "own" the productivity improvement techniques needed for the implementation of productivity improvement actions. Productive skills that are necessary to be acquired by all operating employees are then to be passed on by the managers and supervisors, with the help of consultants if necessary.

Productivity training must cover two major areas, namely:

1. The meaning, significance, and methodology of productivity proper

2. The tools and techniques that can be used to improve productivity

The first area of productivity training topics includes the concepts of productivity, their measurement, analysis, and interpretation, the factors causing variations in productivity levels, and those that bring about changes in productivity trends over time. The second area includes the vast variety of managerial techniques and methods that have been proven to stimulate productivity improvement.

Productivity teaching and training must begin with the creation of the proper understanding of productivity, which is very often misunderstood or misinterpreted, and an awareness of its vital importance to everyone, including entrepreneurs, managers, professionals, and other workers, as well as consumers.

The most important lessons in the field of productivity that we have learned around the world include the following:

- Productivity improvement is vitally important for all of us in today's rapidly changing and highly competitive world.

- Very few people understand what productivity is and how it can be measured and improved, although there is overwhelming evidence that productivity can be increased substantially.

- Personal contact between the productivity specialist and the business managers seems most valuable in bringing about productivity improvement. This highlights the role of the consultant. Even qualified consultants, however, need training in productivity so that they can convey the essential elements of productivity improvement to the business managers in practical business terms.

According to many of the reports received from companies in our interfirm productivity comparisons, *annual* productivity improvements of up to five percent at the company level are quite realistic and sustainable. Many companies have achieved much more. Very substantial variations—up to 100 percent or more—have also been registered among the thousands of competing and comparable companies that participated in the interfirm productivity comparisons.

In specific quality effectiveness programs the concept of quality needs to be conveyed in simple terms. Its vital importance to the survival of the organization, as well as to that of the jobs of its employees, must be explained. The training should emphasize the necessity of incessantly seeking improved productivity and quality, both in processes and outcomes. The training should help all employees understand that their organization exists to serve and satisfy its customers' needs and wants. It is also necessary to explain how to deal with customer feedback. All employees must learn that customers now demand quality products and services at acceptable prices. Otherwise they go to another supplier and the employees' jobs disappear.

Productivity and quality are enhanced today by a trend toward skill diversification. Employees able to handle a variety of related issues and jobs are more useful to the organization. They can also handle, for example, a variety of services for the customer. It has been observed that customers tend to be more comfortable and satisfied with a service if they can deal with one employee or not more than a very few employees in all aspects of the service in question. Employers should, therefore, encourage employees to become multiskilled, for example, through various rewards. Some companies now pay their employees according to the number of skills they can handle. The pay of multiskilled employees in such organizations is higher than those of single-skilled workers in the same jobs. It should not be forgotten both that many—if not most—employees are more satisfied in jobs that give them a variety of tasks to do, and more-satisfied employees do a better job.

Supplier involvement in quality assurance and quality training is essential. As much as 30 percent or even a larger proportion of poor product quality can be attributed to defective materials or components purchased from outside vendors. Supplier involvement, joint planning, and close communication will greatly help ensure the quality of raw materials, designs, parts, and processes. Supplier training can ensure that quality characteristics are built into the incoming materials and parts. If satisfactory control charts come along with the purchased materials and parts, inspection of incoming materials and parts can be reduced to routine observations for identification. If suppliers do not have sufficient quality improvement and statistical control programs in place, it is advisable to demand these and help them learn, as well as implement, such methods.

There are many possible corrective activities that can be taken to ensure and improve productivity and quality through training and retraining. To sum up some of the highlights:

- Training and information sharing that lead to increased organizationwide productivity and quality awareness and involvement

- The tightening of procurement specifications

- Increases in quality standards

- Improved design

- More systematic and purposeful organization and production planning and scheduling

- Improved rearrangement of workloads

- Introduction of up-to-date technology and acquisition of better equipment, when needed

- Improved maintenance to reduce substandard production

- Detailed data gathering and the use of statistical quality control techniques

- Improved working conditions, better personnel policies and labor relations, quality rewards and incentives

The productivity concepts also have already been discussed in some detail. It is necessary, however, to briefly review the relevant basic considerations. First of all, it is essential to teach that productivity is a relative concept. It does not mean much to say that one's productivity is so much or so much, without also saying "compared to what?" We can look at *productivity at one moment in time* to see whether we are better or worse than somebody else. In this case, we are talking about a *productivity level.* On the other hand, when we are analyzing *productivity change over time*, we are making the *comparison with a base period* and call the movement a *productivity trend.* The two productivity concepts, that is, *levels* and *trends*, need to be defined clearly and distinguished.

Before something can be done to improve productivity, it must be measured in order to see where we are, how we are doing, and why. Measurement and analysis are the keys to productivity improvement because the facts must be known before effective corrective action can be taken. Productivity measurement has also been found by companies to be an excellent training technique in itself, not only for productivity but for better business in general, because it forces clarification of outputs and inputs.

The basic elements in productivity measurement training include the identification, quantification, and measurement of corresponding outputs and inputs; their weighting together to make different outputs or inputs, respectively, additive; the method of deflation or indexing to express outputs and inputs in volume terms (excluding the distortions by inflation and other financial factors); making adjustments for variations and changes in the quality of outputs and inputs; and matching inputs with the corresponding outputs.

Productivity measures are different from conventional accounting but they can and should be coordinated. The data need to be accurate, simple, understandable, consistent, and meaningful to the organization. They should have a large degree of acceptance and support at all levels in the organization. It is important to keep in mind that poor measures are not better than no measures because they could be misleading. The measures need to be analyzed in a systematic and comprehensive framework in order to reveal mutually reinforcing or offsetting findings and to eliminate actions that might cancel out each other's benefits. The degree of detail to be taught will, of course, have to vary depending on the responsibilities of the various employees. A basic understanding of the productivity concepts and measures, however, needs to be taught throughout the whole organization.

Among the requirements for managerial productivity training, six functional areas stand out, namely, planning and forecasting, organization, motivation, measuring, control, and marketing. Good written plans are necessary for achieving higher productivity and better overall business performance. They need to be implemented through systematic, properly controlled procedures. Good plans balance and optimize the use of inter-related elements, for example, labor and equipment per capita, or desirable output and marketing levels. They identify and eliminate unnecessary work that can be avoided without jeopardizing the sale of the product or service. Good plans minimize the avoidable interruptions of work caused, for example, by unnecessary engineering change orders, schedule changes, or volume variations.

In the field of resource utilization it is particularly important to teach managers the proper use of time. Avoiding unnecessary chores permits the managers to use their time for their main functions of planning, organizing, directing, guiding, and controlling. In the field of production, the principles of design improvement, specialization, simplification, standardization, and variety optimization seem to be important topics to teach. The quality of output is becoming recognized as more and more important, not only because higher quality results in reducing waste, scrap, reworking time, engineering and management time, and warranty problems, but also because quality output increases customer satisfaction. It seems justified to emphasize repeatedly that higher quality results in strengthened competitiveness. Higher quality is often much more important than a better price.

Choosing the right training program is very important. Feedback and post-program evaluation help improve the training programs and make them more effective. The evaluation should involve the participants, trainers, entrepreneurs, trade unions, the academic community, as well as the various relevant levels of government, if applicable.

Evaluation forms may include the following useful questions:

- Was the training session useful in meeting its objectives?

- Was the information in the handouts informative?

- Were the presentations:

    a. Useful to you?

    b. Adequate in the time allotted?

    c. Appropriate in the choice of topics?

- Was there opportunity to participate and express your views or make suggestions for additional topics to be covered?

One point that still needs to be stressed in training is that while changes do occur and need to be faced, changes cause problems, interruptions, and other costs to business, and should be made only if they are necessary and worthwhile.

    In conclusion, it must be emphasized that productivity training is not a one-time, one-shot exercise. In view of the importance of productivity for business success, it seems essential to introduce the concepts of productivity, performance value, and the importance of quality into the educational system. Further relevant training, retraining, and development need then be provided by employers on a continuing basis. The integration of basic education and marketable skills is needed, and close cooperation between schools and business is recommended.

    In order to be effective, productivity training and retraining need to continue as long as society and the economy keep changing. Industry associations can assist in the productivity of their members. They can define the critical success factors of their industry, develop productivity and quality standards, initiate analytical interfirm productivity and quality comparisons, organize appropriate training seminars, prepare productivity and quality training manuals for their industry, as well as provide prizes for industry productivity and quality leaders.

# EFFECTIVE TRAINING METHODS

There are many ways to achieve the training and development objectives, including:

- On-the-job training under close supervision
- Special seminars or courses
- Study groups, quality circles, or productivity improvement teams
- Evening or correspondence courses
- Programmed instruction, case studies
- Facilitated discussions
- Conferences and lectures

- Encouraging employees to upgrade their skills

- Provision of administrative leave for training purposes

- Directed readership and manuals

- Films, specially designed audiovisual presentations, and tapes

- TV, Internet, or multimedia lectures

- Special work assignments

- Appointments to interorganizational teams for complex problems

- Job rotation

- Use of consultants for training

As to training methods, organizations apply a variety of approaches suited to their individual objectives and environments. During orientation, the manager or an experienced employee who is good at teaching and training should introduce the new employee to fellow workers and take him or her on a guided tour and orientation of the facilities, explaining the relevant company policies, procedures, habits, and typical problems. It is useful to develop orientation manuals and checklists, if possible, for each of the main operations, for example, in hotels for accommodations, food, beverage, and other services. The manual should briefly outline wages and fringe benefits, working hours, working conditions, company and departmental policies, methods, and practices. Checklists help remind employees of the appropriate procedures and practices. The customer's interests need to be highlighted, emphasizing the need for courtesy and pleasant and helpful behavior.

Programmed instruction has been found to be the most effective training method for knowledge acquisition and retention. Case studies, business games, and role playing seem to be the best training methods in problem-solving and decision-making skills. Sensitivity training and role playing are considered as the most successful approaches to changing attitudes and for sharpening interpersonal skills. Quality checklists and packaged lists of quality characteristics, prepared by experienced employees, are helpful. Half-hour to two-hour videotapes, DVDs, films, or multimedia developed for productivity and quality training on the basis of actual company experiences have been found to be very effective as introductory orientation training to these vital topics.

Managers and owner/managers of smaller operations can hardly be away for extended courses. For them it is desirable to provide several short courses at convenient locations. Managers can also benefit from visiting other companies to study their operations and from attendance at relevant conventions.

As our changing society has created an urgent need for increased productivity, and as training needs have shifted in importance from technical toward human aspects, the way in which training is delivered has also shown substantial changes. The trend has been to move toward more results-oriented, tailor-made courses in order to meet the specific demands of each organization. There also has been a tendency in the direction of more frequent but shorter courses, modular courses that can be fitted better into work schedules, and more individualized training and personal guidance. The methods are chosen according to the organization's training needs whether the training objective is orientation, knowledge acquisition, or retention in professional or technical fields, management development, or behavioral elements, such as interpersonal skills and changing attitudes.

There is, of course, a variety of formal courses held in training institutions as well as in-plant. Besides formal training courses, face-to-face contact is a very effective approach to introducing change. Examples are practical in-plant demonstrations of the effectiveness of productivity techniques, field projects that enable participants to translate into practice what they have learned in class, workshops, or small-group activities.

Special events are also used for training in productivity awareness, conveying basic knowledge and information, as well as to stimulate action. These include productivity years, months, weeks, or days, productivity campaigns, congresses, improvement suggestions, contests, and rewards, essay contests, stands at trade fairs, and permanent or mobile exhibitions in schools, shopping centers, or community centers.

Reading material is available to those who are interested and willing to learn through reading. Reading is an especially useful and cost-effective method of learning. Productivity centers, therefore, include this medium among their tools for spreading productivity knowledge. Besides books, journals, and news bulletins, productivity posters, brochures, and other promotional publications have also been popular as productivity motivators and conveyers of basic productivity information.

Computers are valuable tools of productivity improvement. Their use is an important aspect of training and a source of information. There is a fantastic variety of information on the Internet. Combining computers with the very complex process of education, however, is not as straightforward as people once thought. Attempting to train students with screens full of text, for example, can be boring and ineffective. Rather, computerized business games and interactive computer instruction with entertaining scenarios can be useful and very engaging.

# THE ROLE OF PRODUCTIVITY CONSULTANTS

Outside information needs to be searched for in order to keep up with developments, including new managerial and technical advances. New methods and techniques can be learned from formal sources, such as seminars, and also from equipment suppliers, media, new staff, and so on. An important source is the knowledge of expert consultants. In most cases it is useless to ask managers what their productivity problem is. The managers usually are not aware of it unless they are in a large firm that has at least one productivity specialist on staff. This is where the role of the productivity consultant or adviser emerges. In large firms, consultants can help define and set up a productivity program and help to recruit staff for the focal productivity unit.

In the case of small or medium-size establishments, the consultant can provide the necessary specialist services at the beginning of the program, as well as from time to time afterward, either on a regular or part-time basis or as the need arises. When a small firm feels competitive pressure and wants to become more competitive, it is usually advisable to call in a productivity consultant right away, before spending large sums on, say, high-tech investments, which are often considered incorrectly to be the solution to productivity problems. The consultant will take a systematic look at all aspects of the operation and the environment in which it operates.

Consultants provide expert information and make relevant recommendations. They explain what productivity is and what types of information and data are needed for making the right decisions to tackle the problems faced. They identify, diagnose, measure, and analyze, for example, the problems of policy, organization, procedures, techniques, and methods. These problems may lie within or outside the organization. Illustrations of outside problems are the uncertain availability of necessary supplies, the possibility of legal restraints, or the appropriateness of the distribution system.

Consultants may serve as *resource consultants*, who are experts in the specific industry in question, or as more generalist *process consultants*, who know how to go about measuring, analyzing, and improving productivity in any industry, or the consultant may provide both types of services.

*Resource consultants* are experts in the specific industry to which the organization belongs. They are expected to provide the professional knowledge and skills relevant to the productivity problems, which they gained from the experiences of successful organizations or experiments in the organization's industry. These consultants point out the pitfalls to avoid in the organization's type of work. The consultants will locate and transfer the required information or data to the entrepreneurs. The resource consultant's diagnosis may be aimed at correcting existing problems or may be progressive and creative in nature, endeavoring to identify present or future opportunities for productivity improvement, as well as threats that may endanger competitiveness in their specific industry.

*Process consultants* are more general consultants and work with the organization in establishing procedures to solve productivity problems. They help managers with the procedures of how to develop productivity measures and how to analyze and interpret the data. They recommend appropriate actions and help determine what needs to be done immediately and what can be done later. In this decision, the consultant considers the availability of resources with attention to the respective costs and benefits. In many cases the consultant is asked to help with the implementation of the recommended actions. The consultant will also frequently be asked to analyze the company's training needs, provide the company's necessary productivity training, and help the company help itself, for example, by showing how to analyze and interpret the data.

It is important for a productivity consultant to be independent. The consultant's advice will be much more objective and unbiased, for instance, if his or her fees are earned exclusively for giving advice rather than also from, say, commissions for selling computers or other equipment. The studies undertaken by the productivity consultant are usually discussed at a very important meeting with top management. This follow-up meeting is greatly appreciated by the companies' management because—as they have often said themselves—the consultants are there to explain and interpret the findings, "hold their hands," and guide them toward the indicated solutions. The consultants should be available to and brought in by the organization from time to time to review and evaluate progress resulting from the productivity diagnosis and action, as well as to discuss subsequent productivity improvement actions.

# QUESTIONS

Q: 11-1  How does an employee know what to do?

Q: 11-2  What needs are to be included in productivity training?

Q: 11-3  What should training in quality encompass?

Q: 11-4  What are the major managerial improvement opportunities
in the field of productivity?

Q: 11-5 How could one delegate more of the functions to subordinates?

Q: 11-6 How do you determine whether your training has been effective?

# SUGGESTED ANSWERS

A: 11-1 Employees will know what to do if they are told clearly and simply what they are to do. They should be informed about the required standards and specifications. The provision of written instructions is preferable.

A: 11-2 Productivity training must include the understanding of the meaning and concepts of productivity, as well as the methods and techniques that can meet the productivity improvement needs.

A: 11-3 Training in quality needs to include both the tools and skills relating to the job as well as the necessity of working and cooperating with the other workers in order to meet the customer's perception of product and service quality.

A: 11-4 Major areas of managerial productivity improvement opportunities include:

- More complete information evaluation

- Improved planning, organization, motivation, control, and marketing

- Improving manpower assessment

- More attention to human relations

- Limiting spending and better utilization of facilities

- More attention to providing high-quality services

A: 11-5 More of the functions can be delegated to subordinates if one trains them to do these functions, provided that they are capable to perform the function but are currently not sufficiently skilled. Specialists could also be employed to train them. More-complex functions could be broken down into simpler elements, which then could be delegated.

A: 11-6 The effectiveness of training can be determined by measuring the performance results and comparing them to the objectives. Investigate whether or not the training objectives have been accomplished. Ascertain what if any changes are needed to the training.

# Part IV

---

# How to Improve Productivity through Organization and Technology

# 12

# You Can't Succeed without a Plan

## THE NEED FOR SETTING OBJECTIVES AND FORMULATING PRODUCTIVE STRATEGIC AND OPERATING PLANS

The first prerequisite of successful productivity improvement is a careful, thorough, and systematic planning of all operations. The overall planning should include explicit and specific plans for productivity improvement as an integral part of the overall business plan. For example, a good plan will achieve higher productivity and profits by identifying and eliminating useless work, spotting work that does not add value to the final product or service, and identifying work that could be combined with other work in order to save efforts and resources.

Everyone needs a plan—organizations as well as individuals. A plan is necessary to indicate how one will achieve one's goals. It is needed to provide firm guidance about how to achieve progress, maximize the output of products or services, minimize the costs, and ensure that the various parts of the organization do not take counterproductive actions. A plan is needed in order to show employees that they may have opportunity for advancement. For individuals, planning is very important to ensure that they "do the right thing and only the right thing," "the right way," and "at the right time." A good and productive plan enables one to make sure that all efforts and expenses are likely to pay off and that waste is reduced to the absolute minimum possible.

Planning is essentially a systematic approach to decision making and managing in general. Management needs to determine what it wants to accomplish, design the best course of action, develop a detailed plan to achieve its goals, and then proceed to implement the designed actions. The plans must also include provisions for future expansion, particularly for a possibly rapid growth of sales. Many companies fail because they did not plan for growth and unexpected demand develops.

Productivity and profitability of an organization require that everyone in the organization prepare a plan. The various levels of management have different roles in the planning and operational process. Each level needs to understand and take into account the roles of the others and devise procedures for backing up one another when the need arises.

The executives of the organization must look several years ahead to set overall goals and basic policies. They lay these out in the strategic plan. Middle managers make decisions at their own level, and in order to enable them to make these decisions they need to know the strategic orientation of the entire organization. They have a shorter time span, one or two years, but need to develop much greater detail in their operating plans, with milestones by which the various phases of operations need to be completed. Supervisors set up detailed plans and schedules for the coming months and develop weekly and daily schedules. The individual factory or office worker or operator has to plan daily.

Strategic planning by the executives of the organization is vitally important for the survival and success of every organization, regardless of size. It requires decisions for the foreseeable future on such matters as what goods to produce or what services to provide from what resources. A thorough strategic plan focuses management's attention on productivity and leads to the identification and elimination of counterproductive factors and actions. There must be a balance in the organization, for example, between "output producers" and "supporters." In a review of their operation, one of the world's largest corporations has found, for instance, that only 40 percent of its employees were engaged in output production while 60 percent were administrative supporters. In their revised plan they reversed the situation in short order, with fantastic gains in productivity and profits.

Fact-based, good strategic and operational plans do not only improve productivity and operational success but are also necessary for receiving support for the operation and raising the capital for business ventures. Lenders want to be sure that the borrower understands the business, has the required business skills, has done enough preparatory work to ensure success, and that the business will generate enough net revenue in time to service the debt. A plan is needed even for choosing a location and renting a space. The owners of premises want to be sure that they won't suffer losses if the space is vacant between tenants and when re-renting the place.

It is also useful, particularly for new or smaller enterprises, to attach to the business plan resumes and outlines of the personal financial situation of the owner and senior managers. They need to show that these executives are committed to the success of the enterprise, have a financial stake in it, and how the business would continue to operate if the owner or manager became incapacitated or died.

With regard to the production of goods or provision of services, the strategic plan needs to set out how the organization's facilities will be employed. Estimates will have to show the expected manpower requirements and what kind of personnel programs will be designed regarding recruitment, compensation, incentives, training, and so on. In the area of marketing, the plan must detail what marketing and sales initiatives will be taken, including defining the target market and how the wants and needs of the target market will be met in order to achieve the marketing and overall objectives of the organization.

It is also important who prepares the plan. Top executives should join representatives of the various functional departments in the planning process, especially in approving the final plan proposal. This will ensure that the leaders of the various functional departments are personally committed to the implementation of their part of the plan. The desirable length of the operating plan and the related budget depends on the type of business, but it should include enough detail to ensure sufficient control of the operation. A clearly defined productivity plan should be an integral part of the overall operating plan.

Forecasts of the products made or services provided, as well as the expected return on investment, should be included in the operating plan. The operating plan needs to describe the policies to be followed as well as the methods of action selected by management or required by law. The operating plan should define where action is required, by whom, and what procedures should be followed, including the standard procedures that should be used in performing repetitive operations. Alternative courses of action should be examined, and the time to be taken for each program activity should be specified in the plan.

Milestones or checkpoints by which certain steps need to be completed should be set. The milestones should not be too close together. On the other hand, in many cases it is advisable to break down longer actions into several distinct elements in order to be able to clearly delineate whose responsibility it is to accomplish a particular objective or its parts. The plan should include a procedure on how to go about changing the plan, if needed. It should be kept in mind that engineering and other changes demanded in the plan and made in the course of production reduce productivity and increase cost. Changes, therefore, should be made only if they are really needed and justified.

It is very important to involve not only the executives, middle managers, and appropriate supervisors in developing the operating plan, but also the workers who implement the plan, or at least their representatives. Employees are more likely to do their best to implement the plan, including its productivity objectives, if they were involved in the plan development. The cooperation of suppliers and other relevant organizations should also be ensured in developing the plan.

In order to be effective, each plan must be "recorded" and then shared accordingly. This means that every employee knows what he or she should do, when, with what methods or procedures, and how completion should be reported. Reminders and feedback are needed at every critical point or deadline. Reports on meeting these deadlines need to reach the supervisor in time for corrective actions that might be needed. If the required reports are late, automatic follow-up action needs to be taken.

# KEY CHARACTERISTICS OF SUCCESSFUL PLANS

Successful plans of businesses, as well as organizations in the nonprofit sector and public service, have tended to show certain typical characteristics. The plan must be systematic, integrating the goals with the budget, and provide for the balanced utilization of available human and physical resources. The plan should ensure a satisfactory rate of return on total assets. In order to achieve this, the organization will have to minimize its unused or underutilized resources and may have to adjust its product mix. It has to harmonize purchasing, production, inventories, and marketing. The optimal time-phasing of all important elements of a plan is also important because delays in one part may have adverse impact on others. Throughout the planning process, successful organizations always consider the impact of their decisions on the organization's productivity.

The plan needs to be completed before production starts. It must be prepared in writing. This is essential in order to ensure that important elements are not forgotten or neglected. It is also necessary for clarifying to all employees the objectives of their own unit, as well as that of the entire organization, and what they need to do to meet these objectives. The inexistence of a written plan tends to lead to piecemeal procedures, underutilized resources, incompatible or unnecessary actions, and low productivity.

The plan should be comprehensive and reflect routine tasks as well as special projects. It needs to be thorough and reflect the significant responsibilities but should avoid unnecessary items and excessive detail. It needs to be realistic, ensuring that the current resources are sufficient for implementing the plan and policies. The objectives should be set out in small enough, identifiable tasks so that their performance can be identified, measured, and evaluated. The plan should be measurable in terms of quantity, quality, cost, and implementation time.

Problems do occur, and changes in the plan may be required. These should be anticipated, and corrective action plans should be designed for their solution. Planning should be repeated at least annually and provide for revisions of the plan if the markets change. If any of the planned actions have already been tried and were unsuccessful, it needs to be determined why they failed, for instance, due to poor planning or being overstaffed or underfunded. Corrective changes should then be made in the plan.

# WHAT SHOULD BE INCLUDED IN A BUSINESS PLAN AND HOW TO DEVELOP A PLAN

A good business plan is needed to ensure the clear definition of objectives and a careful implementation of productive procedures that lead to operational productivity and success. A well-organized business plan is not only a basic prerequisite of the survival and success of a business, but for the outside stakeholders it reflects well on the management team.

There is no specific design for a good business plan, as there is a fantastic variety of businesses. The organization's history, its present, as well as future projections need to be described in a clear and concise way. The elements that tend to be common to most successful business plans include:

- Company name and location
- Describe and evaluate the tentative objectives
- Gather and analyze information on the past and present
- Define performance and its key factors
- List and evaluate the difficulties
- Determine which are the organization's main strengths and weaknesses: manpower, technology, processes, and so on
- Analyze the external environmental factors that influence the performance of the organization
- Forecast the future
- Develop strategies that must be decided by top management

- Develop the operating plan and detailed subplans in writing, as well as the best alternatives

- Evaluate progress regularly and make adjustments as required

The greatest productivity improvement tends to occur if an integrated approach is taken in the planning process, that is, when management designs the various necessary actions in a well-coordinated framework. The need for high quality and the continuing consideration of relevant costs should always be kept in mind.

The various objectives concerning policies, organization, human resources, incentives, communication, training, safety programs, plant and office facilities, technology, environment, production and/or service processes, standards, inventories, and methods of distribution need to be set and implemented in harmony to be effective. Attention must be paid to keep a balance between detail and the extra costs involved in the additional detail.

The objectives can be achieved through a variety of subplans, depending on the nature, size, and complexity of an organization, for example:

- Organization plan

- Management plan

- Financial plan

- Capital investment plan (both fixed and current assets)

- Personnel and training plan

- Sourcing and purchasing plan

- Production or service provision plan

- Capacity plan

- Productivity plan

- Quality plan

- Marketing plan

- Distribution plan

- Implementation plan

The inclusion of subplans pushes down the responsibility for productivity and profits to lower levels. The subplans should be integrated by senior management into an overall plan in order to ensure, for instance, that the plans and production schedules of the production department are in step with the plans of the purchasing or sales departments, respectively, and so on.

In new organizations, the needs may be difficult to ascertain with any certainty before some experience has been built up. In such cases it may be desirable to rent equipment, particularly in the case of expensive, high-tech equipment. The advantages and disadvantages of "buy" versus "lease" solutions should be considered. A few months of experience is usually sufficient to know which equipment is the right equipment to perform a certain task for an organization. It is often desirable, however, to avoid long-term commitments in order to retain the flexibility to choose the optimal equipment. In new businesses even the purchase of used furniture and equipment should be considered in order to minimize cash outlay.

The formal business plan should indicate how much money is available for equipment, furniture, and fixtures. Certain pieces of new equipment—and the staff necessary to operate them—may be postponed by using outside professional services rather than setting up an in-house department and acquiring the necessary equipment. If certain types of services are not required regularly or sufficiently to utilize full-time internal staff, such services can also be obtained from the outside, as needed. Such services tend to vary a great deal in quality and price and, therefore, provision needs to be made in the plan for shopping around for such services. It is also important to provide in the business plan for the proper utilization of fixed assets and working capital. Too often it is not realized how significant an impact inventory and accounts receivable policies have on the productivity and profitability of the business.

The productivity and quality plans should show past and present performance, as well as relevant forecasts. The major factors influencing the productivity and quality of the operation should be listed and their relative importance evaluated. The criteria of quality and their measures must be defined right at the outset, agreed on by all concerned, accurately monitored, and the results fed back to the interested parties.

Capacity planning is part of productivity improvement. It should show:

- How much labor is required to get the job done, how long it takes to do it, and how often it is to be done.

- The separation of all work into fixed and variable types. Fixed work has a known volume and frequency, such as writing weekly or monthly reports. The fixed work is the base level of staffing in the personnel plan. The volume and frequency of variable work are unknown. All additional staffing above the base level is dependent on the volume and timing of variable work.

The major products and services, as well as their respective percentages, should be determined. Similar products and/or services can often be grouped. The organization's most profitable products and services should be determined and their production increased in comparison with the less profitable ones. Price ranges and quality characteristics of major products should be shown. The sourcing of raw materials and components should be indicated. The proportion of ready-made products bought for resale and that of components bought should also be shown in the capacity plan.

A personnel subplan is also required, describing the management and workforce in such terms as:

- Availability of the required skills.

- The capabilities of personnel. A skill balance is needed in the management team to cover hiring, firing, motivating, training, and so on.

- Recruiting, turnover, absenteeism, quality of work life provided, unionization, as well as personnel problems.

- The composition of the various elements of recompense, including wage rates, incentives, paid holidays and vacation, pensions plan, medical and dental benefits, working hours, method of keeping time records, overtime rules, days off, transfer and separation rules, appraisal system, as well as grievance procedures. Nowadays there seems to be a strong tendency to place more emphasis on rewards for performance, that is, the variable portion of compensation, and making the variable portion of the compensation really significant rather than just a token. It is important to note that good wages and salaries are positively correlated with high labor productivity.

- The costs of infrastructure needed for personnel and the paperwork necessary to provide all the personnel requirements.

- Availability of training, its methods, and duration need to be covered. These should take into consideration the interests of the employees, as well as those of the organization.

The marketing subplan is one of the most important elements of the overall business plan. It should include realistic sales and market share targets, as well as the strategies and tactics needed to achieve these objectives.

It is necessary to ensure that there is sufficient demand in the market for your products and services and that there is opportunity for growth. Additionally, the plan should show what the customers' expectations are concerning the characteristics, quality, and timeliness of the products or services.

In the distribution plan it must be considered what your distribution needs are in terms of:

- The type of product to be distributed

- Frequency of need

- Fluctuation of demand

- Location of customers

- Relative alternative costs of your own or commercial distributors

The overall plan should include a management and implementation action plan, including major opportunities for productivity improvement. In developing the plan, the following points need to be taken into consideration:

- Past experience of potential pitfalls in the business.

- The relevant regulations, red tape, and their possible solutions.

- The outside factors affecting the organization.

- Standards need to be set, against which performance can be judged, for example, productivity-type ratios and preferred financial ratios.

For most organizations, a business plan of 20 to 25 pages is sufficient, but a brief summary should be included. It is important to ensure, however, that assumptions, forecasts, and calculations are correct and realistic. In order to meet this requirement, particularly for smaller organizations, it is useful to seek as much outside input as possible, including that of management consultants and accountants.

Information needed for planning is available from many internal and external data sources. These are discussed in several parts of this book. Some of the required information may be readily available on the Internet. It is important, however, that the so-called "facts" should be verified before they are accepted.

# SCHEDULING

In order to achieve the objectives of the strategic and operating plans, immediate plans need to be developed with detailed operating tactics in the form of easily understood, time-based daily, weekly, and monthly schedules. Scheduling is essential for project management. It is needed in order to ensure that the project is completed on time and within budget. It is useful for revealing unnecessary steps, duplications, and backtracking. Priority criteria can be set so that the most needed items are produced first. Activities that are especially time-consuming can be identified, their value added and cost estimated, and attention focused on the greatest problems. Capacity capabilities can be maximized by setting the stable production volume first and absorbing production peaks through casual labor, overtime, or outsourcing.

Personnel schedules are required for scheduling and monitoring staff requirements for production, hiring and laying off staff, pay and compensation packages, as well as training. Proper scheduling needs to ensure the availability and use of the required worker expertise at the right time. The marketing schedule sets out the tactics of the marketing strategy, such as the development and placement of advertising, sales campaigns, and sales monitoring.

In most operations it is virtually impossible for managers and supervisors to remember every detail and, therefore, it is useful to develop planning and scheduling devices, such as charts. The charts help in controlling the project because they can show progress as related to the plan. By constructing such a network the project manager is forced to consider each activity, whether it can only be performed after another activity is completed, or whether it can run simultaneously. The charts help managers avoid unrealistic schedule expectations.

A huge volume of appropriate project management software is available on the Internet for optimal scheduling activities, as well as for assigning, costing, and tracking human and physical resources. Many Web sites explain the various scheduling and charting systems. At the time of writing, more than 300 Web sites contained the words "Gantt chart systems," including examples of how to draw and use a Gantt chart and its alternatives. We found 1.3 million Web sites on "network management" techniques, as well as a similar number probably including both program evaluation and review technique (PERT) and the critical path method (CPM). More than 200,000 Web sites refer to CPM alone. Some of the Web sites offer example charts, spreadsheets, and software to determine, design, and prepare the critical path, but be aware of possible Internet user fees.

Every activity needs some time, labor, and other resources and has to be included in the budget. Every project needs to have a work order or a memorandum with a serial work order number that describes the work that is to be done, the activities to be scheduled, as well as the required physical and personnel resources that have to be allocated. The serial number is necessary to control that project and has to be placed on all documents related to that project. The work order or memorandum sets out the desired production completion and delivery dates and times. It initiates purchasing materials and other inputs. The orders for materials, components, and other requirements are to be timed properly so they are available for the production workers at the right time. A project can often be completed earlier than originally scheduled if the initial staffing is enhanced or more equipment is allocated.

An effective planning and scheduling chart has three basic requirements, namely, completeness, sufficiency in detail, and clarity. The different methods used are described briefly below in order to explain what is available. Nevertheless, it often is desirable to utilize the services of an experienced consultant to design the most appropriate scheduling method for a particular type of work.

The most popular scheduling techniques are the Gantt chart and the network management techniques. Figure 12.1 shows a basic *Gantt chart* for activities, persons, and/or machines with the duration of tasks and progress on the scheduled work in a time-related sequence. The Gantt chart is a graphical visualization of the length of time an activity requires and

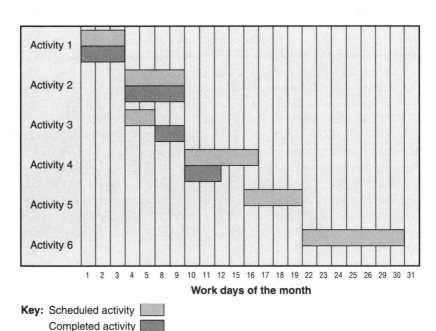

**Figure 12.1** Basic Gantt chart.

whether other related activities are also occurring simultaneously. It can show the entire project or just parts of it, for example, one week at a time.

The Gantt chart system originated around the middle of the twentieth century. It is a bar chart that shows visually when each activity in a project is to start and be completed. The activities are listed on the left in rows. Some tasks can only be started after other activities have been completed, but other activities can run concurrently with others, such as activity 3 in our example Gantt chart.

The allotted times are shown as scheduled start and completion limits on straight lines from left to right, usually daily. The days are shown as columns. The progress of work may be shown in relative accomplishment terms by parallel horizontal lines on the same chart. Gantt charts can show work days only, such as in our example chart, or all calendar days. There are also other charts in the Gantt chart system, such as hourly work scheduling charts for personnel and machine use. There are charts in detailed or summary form.

*Network management techniques,* which are more complex than Gantt charts, consist of flow diagrams showing the planned sequence of activities that need to be accomplished to reach the objectives of the project. The charts show their internal relationships. They enable the consideration of more detail than the Gantt charts, for example, personnel and other resource requirements.

The network management techniques include the PERT and CPM charts. Both consist of a network of nodes and connecting lines or arrows. Each activity follows another one unless it can be done at the same time, and when it is completed rejoins the main path.

When a network is designed, every activity in the project must first be identified. Then it has to be determined which activity must be completed before another one can start, as well as those that can run simultaneously. The start and completion time of each activity is estimated and all the consecutive durations are summed up. Each task has to be costed. The names of persons assigned can be attached to each task. Any milestones such as points of review also can be indicated. The chart can describe the entire project or its various phases.

The path that consists of activities that can only be run after another is completed is the *critical path*. The total project can not be completed by its deadline unless all activities on the critical path are on time. Bottlenecks can be reduced or eliminated, and the entire project speeded up—"crashed"—by adding more staff or other resources at additional cost. If any segments are to be crashed, the cheapest segments on the critical path should be selected to be crashed first. Some segments can not be crashed due to physical constraints, for example a certain time is needed for paint to dry. After such adjustments, another critical path may be generated. The activities that can run concurrently with others are *non-critical.*

# DECISION MAKING AND PROBLEM SOLVING

Everybody involved in the planning and implementation processes needs to make decisions and resolve problems. The decision-making and problem-solving processes are really individualized steps in the overall planning and managing process. The decision making and problem solving must start with clearly defining the question to be decided or problem to be solved in as simple terms as possible. One should consider the question or problem as to its complexity, urgency, resource needs, and availability, as well as its financial implications.

Decisions affecting the entire organization should be made at executive level. Supervisors can be assisted in solving difficult problems by middle managers. In selecting the decision-making level, it is useful to keep in mind that an excessive centralization of decision making may deprive the organization of the flexibility that is needed for quickly adjusting to internal or external supply/demand changes. Decision making should be delegated to the lowest level capable of making the decision. Routine decisions should be made in advance in the form of standard policies and procedures. For some specialized or complex decisions, outside consultants may be brought in to supplement internal capabilities with specialized knowledge or information.

Once the questions to be decided or problems to be solved have been defined, the relevant facts need to be gathered and verified. The facts should be analyzed carefully, taking into account the various interrelationships. Problems should be anticipated and, if possible, resolved in advance. Various possible alternatives should be developed, their respective consequences and risks weighed, and the best alternative chosen. The final decision should provide for follow-up action and the correction of errors as necessary.

# QUESTIONS

Q: 12-1  List some self-analytical questions that can be asked in order to ascertain that adequate plans have been developed for the organization or that the plans can be improved.

Q: 12-2   Why is there need for strategic and operating plans?

Q: 12-3   What are the characteristics of a successful and productive plan?

Q: 12-4   How can the manager of a small or medium-size establishment find time to plan?

Q: 12-5   What does it mean that a plan should be balanced?

Q: 12-6   Why should your objectives be set in a time framework?

Q: 12-7   How can you contend with influences from outside your organization?

# SUGGESTED ANSWERS

A: 12-1   The following are some analytical questions that can be asked in order to ascertain that adequate plans have been developed for the organization or that the plans can be improved:

- Are there clear objectives for the organization, and are the objectives detailed in written operating plans with specific milestones, as well as set against their costs?

- Are these objectives set in a time framework?

- Are the objectives set for small, identifiable groups or manageable segments so that performance can be identified and evaluated?

- Are there written definitions of the organization's overall productivity goals and objectives?

- Do the written objectives and plans contribute to the development of a habit of thinking in terms of productivity improvement and real costs by continually questioning the status quo and asking such questions as "Is this operation needed?" or "Can it be simplified or eliminated?" or "Is there a better way of doing this?"

- Is every employee in the organization working toward quantitative and qualitative objectives?

- Are there specific areas in your organization's operations where additional planning would be beneficial? Which one(s)?

- Can all employees specify what objectives they are working toward?

- Are capital investment decisions, including high-tech equipment purchases, fully justified?

A: 12-2   Strategic and operating plans are needed, for instance:

- To receive support and approval for the program

- To show lenders that they as borrowers have the necessary skills to run a successful business, as reflected by the quality of the plan

- To prove that they have done enough preparatory work to ensure success

- To demonstrate that the business will generate enough net revenue to service and repay the debt

A: 12-3   Successful and productive plans need to be:

- Prepared in writing, systematic, and balanced

- Thorough and comprehensive, and concentrate on the "significant few"

- Reliable and useful

- Measurable and completed before production or service starts

- Aware of the impact of decisions on productivity and profitability

A: 12-4   In order to find time to plan, a smaller entrepreneur has to organize his or her business productively by delegating simpler tasks to other employees so that sufficient time is left for the very important function of planning.

A: 12-5   Balancing a plan means that it should make optimal use of all available human, material, and physical resources. In order to achieve a balanced utilization of capacity, changes may be required to the size of output, product mix, the maintenance of adequate backlog of work, keeping alert for unutilized resources, keeping production, inventories, and marketing in balance. Optimal time phasing of all elements is important because delays in some areas may cause problems in others.

A: 12-6   One has to organize one's objectives in a time framework in order to:

- Get the important actions done in time

- Be able to optimally schedule resources and thereby implement work in a smooth and balanced way

- Make the best use of allotted time

A: 12-7   In order to contend with outside influences, the managers need to:

- Learn the relevant legal and regulatory issues and decide how to cope with them

- Analyze the environmental factors and requirements as they apply to the firm, assess their importance, and determine how to best meet them

- Take action regarding their implementation by assigning responsibilities, arranging priorities, and making arrangements for controls and evaluation

# 13

# How to Design Productive Operations of High Quality

## SETTING PRIORITIES FOR AREAS OF POTENTIAL PRODUCTIVITY IMPROVEMENT

One can't tackle all productivity problems at once. The most important ones must be tackled first. In order to assess the relative importance and priority of the various opportunities for productivity improvement, it is important to weigh them against the following criteria:

1. Is the problem a pervasive, organizationwide problem, or does it only reflect a part or a specific aspect of the organization, such as a department or a function?

2. How great is the relative importance of the problem, say, in terms of its impact or its costs? For instance, a labor-intensive or a high-labor-cost operation would probably require a close look at its labor productivity and its determinants.

3. Is it a basic problem that caused the low productivity, for example, poor materials, lack of skills, employee apathy, or is it a symptom rather than a cause, for instance, high turnover or absenteeism rather than underlying poor labor relations?

4. How soon will the correction of the problem lead to successful productivity improvement? Particularly in the initial stages of a productivity program early success is important in order to strengthen the program's credibility.

5. Can the problem be resolved by known methods or techniques of modern management, or is it not suited to ready-made solutions?

6. Is the problem mainly human–managerial–organizational in nature or is it primarily physical–technological–technical? While the technological aspect tends to receive more public attention, most findings indicate that the human elements of productivity and quality are at least as important as the technical ones—if not more.

7. Are the causes of the problem within or outside the organization? Productivity problems can be caused, for instance, by poor labor relations within the organization or by erratic supply of raw materials needed from the outside.

8. Closely related to the above question is whether the corrective action can be taken within or outside the organization. A productivity study by the International Labour Organization (ILO) confirmed that some 80 percent of productivity improvement actions of enterprises can be taken by managerial action within the organization.

9. Is the product or procedure too complex?

10. Are there unnecessary procedures or services among the activities?

11. Is a simple and clear procedure manual needed?

12. Are the responsibilities and accountabilities clearly assigned?

13. Is there an urgent need for training in certain areas?

The vital importance of quality to productivity must be kept in mind in all productivity improvement endeavours. The question of quality is discussed in detail in the following sections.

# CUSTOMERS AND CLIENTS NOW WANT PROOF OF QUALITY

The importance of quality for productivity and success has become increasingly recognized. In order to satisfy the customer and improve domestic and international competitiveness, the number of permitted errors and faulty products were limited to one percent of output. Subsequently, the drive toward zero defects has been spreading.

By the beginning of the twenty-first century it had become recognized that wider impact on quality improvement can be achieved if not only the best but any high-quality production and service is officially audited and certified by independent bodies. This applies to the public sector as well as to private organizations. Today's customers are not only expecting and demanding a quality product or service, but also demand proof that the organization producing the product or service is consistently capable of producing high-quality products and services. Furthermore, people have more confidence in products and services that meet international standards. The increased demand for quality assurance during all stages of the product or service delivery process (design, manufacturing, performing services, delivery, and so on) has resulted in the need for organizations to work to recognized international standards, hence, the promulgation of ISO 9000 standards. These are standards for high quality that assist the organization in controlling the quality of the activities of their organization from within their organization.

The ISO standards were and are developed by the International Organization for Standardization (ISO) by consensus of the participants.[1] It is a nongovernmental network of national standards bodies from (at the time of writing) 157 countries, one per country, working with governments, industry, business and consumer representatives, as well as other international organizations. It is financed 60 percent by membership fees and 40 percent through sales of publications and other income from services. The ISO was established at an international meeting in London in 1947 to coordinate and unify industrial standards on an international basis. The purpose was to facilitate the exchange of goods and services globally, and to develop international cooperation in the areas of intellectual, scientific, technological, and economic activity. The headquarters of ISO is in Geneva. A typical example of an ISO standard is the size, thickness, and other characteristics of credit cards that are used around the world.

At the time of writing, more than 17,000 ISO standards have been developed. They are periodically reviewed. The ISO standards initially started with technical standards. Then the ISO 9000 quality management standards were developed and have increasingly become the international family of standards for quality requirements in business as well as governments, for production and services. These are generic standards that can be applied to any organization, public and private, large and small, producing goods or services.

The ISO 9000 standards are management system standards and refer to how the organization manages its activities. They deal with the requirements of quality management—what and how an organization does to satisfy its customers and clients. ISO 9000 is a family of standards, which includes guidelines and technical reports relating to quality management. Supporting ISO standards assist in the preparation of quality plans, processes, product accuracy, and the preparation of quality manuals.

At the end of 2007, publications of the ISO by technical section included 295 in engineering technologies, 367 in electronics, information technologies, and telecommunications, as well as 129 from the transport and distribution of goods section.

The ISO principles require that the organization understand and meet the needs and wants of their customers, and that the organization's leaders ensure the unity of the organization for this objective. All staff and all processes should be managed systematically; all operations should be based on factual data and information, and the organization's performance should be continually improved.

The benefits of adopting ISO 9000 include cost savings due to built-in quality at every manufacturing stage, reduced waste that would have been caused by reworking, reduced product liability claims, and a complete record of every stage of production that is invaluable for product and service improvement. The ISO 9000 standards apply many of the principles and requirements of quality that are described in this book.

One of the first large international productivity organizations, the Asian Productivity Organization (APO), organized in 1993 in Hong Kong its first training course on ISO 9000 for the manufacturing sector. The training course was designed for senior managers, government officials, and consultants in APO's member countries concerned with quality improvement in the manufacturing sector. Their purpose was also to assist local companies in obtaining ISO 9000 certificates. Today, ISO 9000 continues to gain acceptance through recognition of quality achievements, and is rapidly becoming the required family of standards for products and services provided in most of world trade.

ISO registration demonstrates that the organization is serious about quality and that top management is committed to the effective functioning of the quality system. Policies and practices are designed to ensure a service-oriented work culture and pursuit of the objective of continual customer or client satisfaction with the products and services provided. Systems and processes are employed that ensure consistency, reliability, and conformity with preestablished norms, values, and methodologies. Internal discipline is introduced and enforced to ensure that the right thing is done the right way the first time. Systematic improvement of performance is ensured by introducing mechanisms that will provide a basis for proper education and training of all staff.

Organizations that are registered and meet the ISO requirements receive a certificate of registration from the accredited certification body or registrar, and are listed in a directory of organizations of demonstrated and independently assessed capability of producing high-quality products or providing high-quality services. In assessing the quality system, assessors or auditors are looking to see that the system is capable of ensuring that the organization's products and services meet specified requirements. The ISO certification provides customers and clients with guaranteed high levels of product and service quality.

# HOW TO ENSURE HIGH-QUALITY PRODUCTION AND SERVICE

The years straddling and following the turn of the twenty-first century have been witnessing a new "industrial revolution" brought about by a new focus on people, quality of production and service, unparalleled technological change, and globalization of world trade. This revolution has affected the service industries as well as goods-producing organizations.

The new organizations can utilize high technology, respond promptly to changing market conditions and demand, as well as create a culture of superior performance throughout all operations of the enterprise. In many operations, such as in just-in-time (JIT) manufacturing, all incoming components must meet very high quality standards so that inspection is not needed. The quality improvement of today's modern enterprises is characterized by frequent product improvements and innovation. Quality improvement is brought about simultaneously with the containment of variable production costs. The fundamental changes have not only stimulated but also enabled the creation of flexible multiproduct firms in place of the predominant mass production enterprises.

These rapid worldwide economic, commercial, and social developments have brought about the need for and concept of total quality management (TQM). This recognizes that weaknesses in any part of the operational process have significant effects on the final output of products and services. Total quality management means, therefore, that all employees need to aim for high quality in all aspects of the organization and that the organization is managed to ensure the satisfaction of the customer or client. It comprises all basic components, namely, product and service quality, process quality, and human quality. Typically, a TQM system leads to a flatter organizational structure in which all management and staff work as a team. Managers become facilitators from bosses, continual training is provided, and performance is regularly measured and evaluated.

It is important to ensure the maintenance of high quality even in the case of stiff price competition that otherwise may lead to indifference to quality and result in lower quality. Lowering quality leads to reduced competitiveness and endangers the survival of the enterprise. Quality improvement can not be brought about from one day to the next. Many months, perhaps even years are typically required to achieve major changes. Quality improvement is not a one-shot exercise, but must become a continuing commitment and endeavor to be successful. Nevertheless, visible success can often be achieved in relatively short periods.

All quality problems can not be tackled together at the same time but, like all productivity problems, need to be prioritized, following a plan to improve quality. The areas that should be addressed first include problems that affect large parts of the organization and its products and services, in particular those that are costly and promise early solution. Nonetheless, it is advisable to keep closely related and affected areas in mind when trying to solve problems.

The new quality strategy usually permeates and brings about fundamental changes in all departments, processes, and jobs. The avenues leading to quality improvement range from corporate strategic planning to process control and error prevention within the enterprise. Management should ensure that quality is no longer limited to maintaining standards, but also includes providing employees with guidance, technical support, and skills to better their own work.

One of the most effective strategies of successful companies that achieve strength is when they measure, analyze, and, as a result, know their quality performance and can correct the identified problems. Unnecessary, redundant, and confusing items of information are to be avoided. Unfortunately, most companies still do not have adequate—or any—data on the quality of their products or services or about the cost of poor quality. Neither do many understand the positive relationship between quality and productivity.

All inputs affect quality, including:

- Management's capability, effort, and concern for quality, as well as the organization's relevant strategies, policies, processes, and specifications

- Knowledge, skills, and effort of labor, as well as their working conditions

- Materials and components, their quality, suitability, and dependability

- Buildings, machinery, and equipment, their suitability, reliability, and maintenance

- Customer feedback

The available data indicate that as much as 80 percent of quality problems are due to the organization, the system, the lack of managerial concern, as well as poor leadership, inadequate design, product engineering, procedures, or quality specifications. Only 20 percent tends to be due to the first-line operators.

The new quality management strategies of successful organizations are characterized by commitment to people and respect for the individual employee. Workers are given greater understanding of why quality changes are needed by their organizations as well as by themselves. They also are provided an opportunity for input into determining and achieving their organization's objectives. As a result, the workers develop a much greater sense of control over what they do and how much they earn. Their ideas on problem solving and quality improvement are increasingly being rewarded.

Quality management is not only necessary at the enterprise level, but must be part of handling all individual projects. The projects need to be defined according to standards based on customer or client requirements. Performance must be measured and controlled to ensure that projects adhere to the standards. Potential deviations need to be forecast, and correcting or at least mitigating strategies should be developed. These are only possible if the standards are realistic, measurable, and withstand the scrutiny of all stakeholders.

For the purpose of addressing these objectives, the role and value of both product and service quality, as well as what employees can personally do to improve quality, must be communicated and explained throughout the entire organization. Understanding helps ensure the voluntary cooperation of all employees. Discussions of quality objectives and problems help ingrain the importance of quality improvement and develop an organization-wide habit of continually seeking better quality. Positive communications concerning quality can be more stimulating than regulations. It is useful, for instance, to tie in quality improvement with personal pride, praise, rewards, or new opportunities rather than enforce it through regulations.

The lack of necessary skills, training, or experience of any employee can obviously cause quality problems. It is equally important that all concerned know that it is essential to do a high-quality job, what to do, and how to do it. Output quality can only be improved if management and all workers know what quality is and how important it is to their customers as well as to the survival and success of their organization and their jobs.

It is important that quality performance objectives guide all personnel operations. Interest in quality performance and quality improvement should be given significant weight in recruitment, selection, compensation, training, and retraining. These should include key processes, such as appraisals, recognition, incentives, and rewards. Priority should be given, for instance, to candidates showing evidence of commitment to quality. In the case of developing compensation packages, considerable weight should be given to the results of quality performance. The quality of work performed by personnel should be characterized by fairness, speed, timeliness, accuracy, and respectability.

Defective materials and poor workmanship of components are effective barriers to better quality and productivity. It is essential, therefore, to ensure the high quality of incoming goods in terms of suitability, reliability, and uniformity. Meeting the prescribed specifications may not be sufficient to ensure their quality unless the specifications have been based on observations during the production processes where the materials or components are utilized. The specifications may have been written by people other than those in production, and the specifications may, therefore, be incomplete and not suited to the requirements of the production process. Purchasing managers must cooperate with the production department as well as suppliers in order to ensure the uniformly high quality of incoming parts and components.

In the area of physical facilities of production, quality problems may be caused by obsolete, unsatisfactory, or poorly maintained equipment and buildings, or overcrowded surroundings, noise, and inadequate lighting or ventilation. Safe, healthy, and pleasant surroundings help improve quality due to more-satisfied employees. Equipment or buildings in need of repair tend to cause breakdowns and lower-quality output. In order to eliminate waste of energy, one should make sure that the heating and cooling equipment operate effectively. In professional activities, effective and well-maintained technical libraries, Internet access, and/or data banks also are very helpful.

The customers and clients themselves are important contributors to the quality of products and services. Since the required quality is determined by them, feedback from customers is essential on what quality they want and what they are willing to pay for it. They can express their needs, wants, and desires regarding the goods and services they buy, suggest problems and possible solutions, express satisfaction with good performance, or criticize unsatisfactory products or services.

# GREAT OPPORTUNITIES IN ADMINISTRATION

Many productivity and quality improvement methods apply to all functions of economic activity, and are discussed throughout this book. The methods of determining the specific needs of productivity and quality improvement were discussed in the analytical Part II of this book. The sections following cover those specific aspects of productivity and quality improvement that are not fully covered in other sections.

It has been observed in a vast number of private and public organizations—small as well as large, even very large units—that administration uses an inordinate portion of resources as opposed to the production of goods or the provision of services. We have to keep in mind that administration includes many such activities required in business as general management and specialized administrative work in activities of finances, planning, personnel, purchasing, production, warehousing, sales, distribution, billing, and collection.

As a consequence, improvements of productivity in administration usually offer great opportunities for enhancing the performance of organizations. Above all, it needs to be ensured that their activities are necessary. Much administrative work may not be needed for the achievement of added value and should be eliminated. It is important, therefore, to know the proportion of productive work. This examination should be done thoroughly and reviewed from time to time.

Apart from productivity improvement actions that have already been discussed, improvements can be brought about through better office procedures, methods, and systems. These consist of a series of logical steps that must be followed to accomplish a particular undertaking. They include collecting facts, performing analyses, developing various alternatives, weighing the probable consequences or each, and then settling on a preferred course of action. Time-consuming procedures can be identified by:

- Clearly defining steps in the process

- Measuring the time consumed by each

- Measuring the value added of each

- Measuring the time consumed by non-value-added steps

- Focusing on the greatest problem

In order to enhance productivity, all activities should be justified in view of their purpose. Questions should always be asked regarding productivity and quality. The relevant facts should be gathered, organized, and evaluated. These should include productivity and quality data for the unit. Red tape, bottlenecks, backtracking, unnecessary actions, waste, and idle time need to be identified and eliminated. Electronic and, if necessary, manual forms should be developed and used whenever feasible. All information and data should be kept in a way to make them easy to retrieve.

The productivity of production management can be improved by:

- Improved planning, reconciling market requirements with the available plant, financial, and personnel resources. Uneconomic product variety and production runs need to be minimized. Products should not be produced in such volume that they can not be sold for more than their full cost. All elements of production should be coordinated to minimize costs and ensure satisfactory delivery of products or services on time.

- Better product design, for example, by introducing screw-less and wireless design, standardization of screws, as well as by minimizing the number of parts and components.

- Improved methods through cost analysis, elimination of redundant work, redundant steps, and unnecessary operations, by simplification, standardization, specialization, sophisticated technology, mechanization, and automation where justified, use of numerically controlled (NC) machine tools, automation of machine adjustments and inspection, fastening by shrink-fitting, or use of sheet metal fabrication to replace the machining of separate parts or components, if preferable.

- Ease of assembly, use of molds to facilitate assembly, reduction of setup time.

- Better production processes by making them simpler, better sequenced, smoother with more levelled production flow, and combining procedural steps where possible.

- More efficient and economical plant layout.

- On-the-job training of operators to improve their technical skills, and laborsaving technology.

- Increasing employee safety by the provision of protective clothing and accessories, as well as by minimizing safety hazards.

- More efficient utilization of materials and energy.

- Calculating whether it is cost-effective to rework rejected products.

- Improved materials handling. An average of about 20 percent of labor cost is an often-quoted estimate of the cost of materials handling. It must be kept in mind that handling does not add anything to the selling value of a product and should be kept to a minimum.

- Improved shop floor data-gathering including online information processing and the recording of production times to make or assemble each component, labor availability, machine availability, percent use of machine capacity, and materials used per component. Relevant records should be kept in order to continually indicate the condition of assets in order to enable better preventive maintenance.

Many of these activities are performed by clerical staff. Their productivity can be enhanced significantly if their work is measured, for instance, by work sampling, and then better work schedules are introduced. Avoidable schedule changes should be eliminated. Clerical work can also be improved by methods improvement and procedural changes. Fluctuations in production volume need to be smoothed out, and production interruptions, equipment downtime, idle time, and wasted efforts should be eliminated.

The distribution of work in the office can be improved if the main activities are defined, if it is determined which jobs take most of the time, and whether the most effort is used for the most important and urgent jobs. Large organizations, such as banks, need to decide what is done centrally and what is delegated to local branches. Coordinated autonomy is today a favored solution, to determine centrally what products and services to sell, how quality is to be ensured, what information technology should be used, and what related earnings are to be achieved. Staff salaries, product prices, and other local branch issues can be decided locally.

It is important to ensure that the available skills are properly used. The jobs themselves can be improved by recording the details of each job, identifying the most important problems in doing the work, choosing and developing the best solutions, and, if necessary, with the approval of the supervisor, implementing the improvements. Besides salary and wages there are other costs including fringe benefits, space and its maintenance, light, heat and air conditioning, computers, telecommunication, as well as furniture. Environmental and safety issues also must be addressed, including work with hazardous materials, performing dangerous operations, and the provision of protective clothing and accessories.

Improved procedures make use of electronic and hard-copy reports and forms designed to eliminate excessive clerical handling, including documentation for various operations, such as order recording, production and service, warehousing, inventory control, transportation, invoicing, and collections. Good office procedures ensure that essential records are protected against loss. All forms, reports, and other clerical products need to be reviewed periodically in order to ensure that they are still worth producing. Superfluous and no longer necessary forms and records should be systematically discarded.

Clerical systems have been increasingly computerized because modern operations handle a large volume of data and are becoming more complex. Effective office systems and procedures must aim for an optimal balance between producing the necessary data and combinations of data at the lowest overall cost. Additions of new electronic data processing (EDP) equipment or hardware need to be examined and justified very carefully in order to ensure that they are necessary and could not be performed satisfactorily with the existing equipment. Data should be accurate, as simple as possible, and self-checking, that is, any errors or contradictions should be automatically evident. The following office operations are particularly suited to EDP application: payroll, production planning and scheduling, inventory control, invoicing and collection, as well as productivity and cost analyses and reports. Difficulties in retrieval must be eliminated.

# PRODUCT DESIGN, STANDARDIZATION, SIMPLIFICATION, AND SPECIALIZATION

Productivity and quality can be considerably improved by better product design. Attention needs to be paid to ease of production, serviceability, cost reduction through the elimination of unnecessary parts and components, the reduction of the number of parts and components, the use of alternative materials, as well as the introduction of new materials if they are helpful to repair or maintenance. Product and process specifications should be prepared and agreed on before the production process is started.

Long-term planning makes easier the development of new products and the upgrading of existing products. Product design and development should be integrated with engineering and production, inventory, and packaging, as well as coordinated with market requirements and suppliers. In order to achieve these goals, suppliers and customers should be involved in the product design process. Both products and services need to be kept under review as to whether they should and could be improved. The value added should be questioned at every stage of production and service. R&D sources outside the organization should be studied, and products and services should be compared with those of competitors. Cost–benefit considerations should be investigated, and the markets should be tested.

Very important tools of productivity improvement in product, service, and procedural design are standardization, simplification, and specialization. What we call "standards" are precise technical and other documented criteria, specifications, rules, characteristics, and guidelines that are to be used consistently. As we have already stated, international standards are based on agreements that ensure that they apply around the world. The application of standards ensures that materials, products, processes, and services are consistent and suitable for their purpose.

Standardization means that standards are applied and that they are applied consistently. It means that many different items made by a company accommodate the use of fewer components than would otherwise be required. As a result, fewer parts need to be purchased, handled, and stored. Standardization enables efficient materials flow and information exchange. Standardization also applies to product and part numbering, computer-based planning and control, operating control, and time management. In view of the importance of standardization, it is necessary to regularly examine all jobs and procedures to determine whether they could be standardized and whether common components, materials, or processes could be used for various products.

Another very effective method of improving productivity and quality is work simplification. Its objectives include the production of better products, or the provision of better services, and the reduction of costs at the same time. For the employees, work simplification can reduce stress and ensure gains in personal income through awards or better promotional possibilities. Offering individual benefits for simplification enhances employee interest in trying to simplify methods and processes and also eliminates possible worker resistance to change.

In order to simplify work, it is necessary to diagnose the work flow step by step, consistently examining at each step why it is done, whether it adds value to the product or service, whether it could be eliminated or combined with other steps, or done in another but simpler way. It is useful to calculate the savings (in both cost and effort) that is achieved by simplification. Opportunities for work simplification can best be spotted by finding repetitive, uneven workload, excessive walking or talking by workers, backtracking, poor layout, and crowded aisles or work stations.

Another tool of better work design is specialization. It is useful because the interests, skills, capabilities, and experience of workers vary, and specialization makes them more productive in certain types of work. Specialization builds experience and makes it possible to produce a product or provide a service better at lower cost than if it were done by unspecialized individual effort. One possible drawback of excessive specialization may be boredom with doing certain tasks over and over. It is, however, possible to find an optimal combination of specialized jobs with diversification into other tasks.

# PURCHASING, PLANT ORGANIZATION, AND INVENTORY MANAGEMENT

The first step of the actual production cycle is purchasing. It mainly includes sourcing, procurement, and cost control. Sourcing consists of finding potential sources of supply, investigating vendor qualifications, checking prices and quality, as well as ensuring the timeliness of supply by various competing vendors. The procedure requires regular reviews of whether it is more advantageous to buy or make the supply, and ensuring the best source. It is often necessary to ask for several bids before purchasing in order to ensure the lowest prices for suitable quality.

Procurement requires choosing the preferred suppliers and negotiating with the vendors to ensure that the supplies meet the required specifications of the organization, including assurance that the suppliers are properly trained and productivity- and quality-conscious. The primary vendors should be integrated into production plans and operations. Productivity and quality of this function can be improved by developing more-standardized purchasing specifications as well as developing and maintaining simple and clear systems for procuring the goods and services required. The specifications can include establishing computerized automatic reordering when a certain level of stock is reached, and reducing paperwork through regular resupply, if this results in overall cost savings.

It also is important to keep in mind that the cost of materials and components is increased further by the cost of receiving, inspecting, handling, and storing the materials, plus any scrap and rework caused by incoming materials. Incorrect deliveries disrupt schedules and also add to the costs. It is necessary, therefore, to regularly follow up on suppliers and subcontractors to ensure proper delivery by the promised date.

The costs of incoming transportation can usually be reduced if:

- Economical quantities are ordered

- Orders are issued in time to allow cheaper transportation

- Route and date are specified

- Incoming invoices and delivery dates are checked against the orders to eliminate unordered goods, unacceptable items, and shortages

- Catalogs and prices of suppliers are kept up to date and checked against the prices on invoices

The purchasing function also includes the continual review of alternative sources, materials, and components. A clear and adequate definition of the work should be ensured on all subcontracted elements. Unnecessary, nonessential regulations and controls need to be identified and reduced, and as much program stability and continuity should be ensured as possible. Vendors, contractors, and subcontractors should be encouraged to improve their productivity and quality to reduce costs.

It is helpful to establish good vendor relations and take advantage of long-term buyer–seller relationships. Cooperation between purchasers and suppliers can be developed through jointly set priorities, lead times, and stocking procedures, as well as through data feedback to vendors on their services. Incentives can be financial or nonfinancial, such as recognition or awards, which may be given for improved quality, lower prices, and reliability of supplies. As much as half of a company's quality nonconformance is often caused by defective material provided by suppliers.

In order to ensure the most efficient and economical plant layout, the areas of production, storage, and handling of products should be arranged in such a way that the production and handling distances and costs are minimized. Machinery and other equipment should be located in an arrangement that minimizes the movement of goods in process and their operators, without backtracking. In stores, cash registers should be placed in central locations and stock areas should be close to the selling areas. Proper store layout is important so that space is utilized effectively. Attractive and impulse items should be placed so that customers notice them.

One should examine whether the location of production and nonproduction departments, such as stores, packaging, and dispatch, cause any problems or offer any opportunities. Nonproduction and nonselling departments should preferably be located in areas that can not be utilized more profitably for selling or customer service. Areas that are less accessible or less desirable to customers should be used for such activities, for example, basements, the back of the store, or upstairs. Lighter materials and components can be moved further easier than heavy material. Therefore heavy material should be placed so that it will require the least movement. The place should not be overcrowded, the aisles should be kept clear, and the overhead space should possibly be utilized.

Proper inventory management is another important element of improving productivity. Productive inventory management requires carefully maintaining the inventories of individual stock keeping units (SKUs) according to the organization's plan. It is important to maintain proper inventory levels. Too low inventories risk lost customers when orders can not be filled when required. Keeping too much inventory and unnecessary duplication must also be avoided because they cause extra cost of idle stock, increase risk of spoilage and obsolescence, and use up valuable space. Private labeling should not be done until orders are received.

Sufficient quantities should be stocked of well-selling items. A full assortment is usually required of seasonal items at the beginning of the season. The orders must be well planned. Work-in-process (WIP) must be coordinated with sales and inventory levels. In order to reduce too large WIP levels, their causes should be addressed. These include the replacement of unreliable suppliers and equipment, or the reduction of too long production setup times.

Inventory levels can be optimized by classifying SKUs by their relative importance in terms of their turnover volume, value, activity, and predictability of use into items A, B, or C. Items in class A demand close control while items in class C need the least. "A" items usually make up less than one-third of all items but tend to account for over three-quarters of total annual sales. "B" items, which include less expensive items of regular use, as well as items that are less likely to deteriorate or become obsolete, account for a large part of all items although they represent only a relatively small fraction of total sales value. "C" items that have low sales may even be made or purchased to order only.

In improving the productivity of inventory management, it is important to keep track of production and sales forecasts, as well as changes in engineering, and then make inventory adjustments as required. Slow-selling stock can be controlled by keeping records of the age of the merchandise. These and obsolete SKUs should be kept under constant review. Inventory reduction may also result in productivity improvement as long as it does not cause a higher cost for meeting delivery deadlines than the savings achieved by reducing the inventory. Reduction can be achieved by chopping no-activity products, using incentives to increase vendor stocking, consolidating warehousing locations, and quoting long lead times for slow-moving and made-to-order items. The productivity of inventory management also can be improved by reducing shrinkage and pilferage through better security.

Orders received need to be recorded, processed, and placed into inventory. Transactions should always be compared with the budget. The growth of inventory needs to be compared with sales growth. Inventory counts should preferably be maintained electronically but the actual inventories should be annually confirmed by physical counts.

In order to improve the productivity of warehousing, which is often estimated at one-fifth of total distribution costs, one needs to keep in mind that warehousing consists of a variety of functions, including receiving, barcoding, and putting away the SKU items into storage. The orders need to be selected, checked, packed or crated, marked, and shipped on time. After the removal of items, storage has to be replenished. The clerical documentation must be completed promptly and accurately without errors in labeling, processing, and billing.

In view of the variety of setups and functions, there are many ways in which warehousing productivity can be improved. Warehouses should be located with regard to suppliers and markets to minimize inbound and outbound transportation costs. The stored SKUs must be kept in an identified place that all staff knows. The SKUs need to be found easily, and fast-moving items should be easily accessible and nearest to the shipping docks. Shipping schedules must be balanced to smooth out peaks and valleys.

The productivity of materials-handling can be improved by avoiding unnecessary movement of materials, as well as unnecessary loading and unloading. Where gravity movement can be used, it can replace horizontal movement. Moving and storing items on pallets enables easy use of forklift trucks. The use of variable-height shelves helps to reduce empty shelf space above the stored product items. More suitable vehicles can enable the use of narrow warehouse aisles.

High-tech equipment is useful in warehousing, if cost-effective, by enabling continuous communication between materials handlers and the office, optical scanning of coded products by electronic reading devices, stamping the storage location on put-away tickets by automatic labelers, as well as generating put-away and pick tickets by computer. Standardized packaging can reduce the variety of shipping cartons.

Smaller individual pieces can be grouped into much larger units for shipment through unitization, resulting in great savings in handling and shipping costs. Stretch- and shrink-wrap packaging improves the productivity of handling and can reduce handling damage. A certain person or persons must be designated to be allowed to enter the warehouse and take parts out. The materials-handling methods must be kept under review in order to ensure that the methods used are the most cost-effective.

# PRODUCTIVITY IMPROVEMENT IN PHYSICAL DISTRIBUTION

Productivity can also be improved in the physical distribution function, which has been estimated to cost as much as 15 to 20 percent of the sales dollar of an average manufacturing company. There are reports that organizations with productivity programs save up to 20 percent of their annual distribution expenses. It has been estimated that nearly one-half of physical distribution costs are for the transportation of goods. The rest is used for administration.

The over-the-road functions of physical distribution consist of loading, line-haul movement from origin to destination, as well as "at stop" activities. These functions include the pre-trip activities and documentation, as well as such end-of-trip activities as checking out and cash settlement. The lack of timely management information and data for comparable standards is among the greatest problems in physical distribution. The parties involved want to know where they stand. The data needed for enabling the development of standards for such matters include driving distances, driving times, and fuel usage. Lack of adequate productivity and other performance data and comparative industry standards for each functional activity are outstanding problems of most companies in physical distribution. Clearly, there is a need for interfirm productivity comparison-type studies to help companies in physical distribution improve their information and results.

The basic organization of physical distribution needs attention from the viewpoint of improving productivity. The size of the operation may need adjustment. For example, studies in certain physical distribution industries, such as carpet distribution, suggested that smaller operations may be more productive and profitable than larger ones because of such factors as flexibility, specialization, and lower overhead costs. Regional distributors may also have advantages over national distributors. The distribution mode mix and delivery methods may also be improved. Routes may be reorganized to minimize the distances driven. Shipments into specific market areas may be pooled and shipped on selected dates. Job descriptions could usefully specify what jobs the operators, such as drivers, are expected to do and by what methods. Driver performance may be improved by better organization such as a driver relay system on regular routes.

In the distributive trades there also are great opportunities for productivity improvements in administration. Paperwork may be reduced by computerization and by soliciting annual service and purchase contracts. Formal quantitative goals should be determined for distribution activities. These should be based on observed and measured facts, preferably in volume terms, rather than based on simple increments of past financial results, which may be misleading under changing circumstances. In many cases, the volume data are already available from order forms and other documentation. This approach may require industrial engineering methods for establishing detailed standards.

Besides the successful productivity improvement opportunities already mentioned, the reduction of empty driving, volume optimization, and improved routing and scheduling offer great opportunities for productivity improvement. These need to be kept under continual review. The route supervisors and dispatchers can identify as well as implement many productivity improvements. It is important, therefore, to hire capable staff and provide them with continual training to help them learn new and better procedures, methods, and techniques. Specialized equipment may be needed that is suited to the type, size, and shape of products to be handled in order to help loading and unloading of vehicles. If transportation is done in the organization's own vehicles, less expensive packing materials may be satisfactory.

# QUESTIONS

Q: 13-1   How can you determine whether your production methods are good?

Q: 13-2   How can you evaluate the quality of customer service?

Q: 13-3   Suggest methods of improving quality.

Q: 13-4   List some examples of important aspects of quality in personnel work.

Q: 13-5   List some of the determinants of choosing productivity improvement actions.

Q: 13-6   Should you maintain buying through established sources or continually search the market for the best merchandise prices and services?

Q: 13-7   Suggest some of the customer's main requirements in physical distribution.

Q: 13-8   How can you reduce accounts receivable?

Q: 13-9   How would you start improving productivity in administration?

Q: 13-10  How can you identify better methods of physical distribution?

# SUGGESTED ANSWERS

A: 13-1   In order to determine whether one's production methods are good, one should consider whether they meet their objectives, produce what they should and how they should, whether they are up to date, and whether there are simpler or cheaper ways of achieving the same objectives.

A: 13-2   In order to evaluate customer service quality, the complaints of customers should be analyzed as to the specific aspects of the problem, and the percent of deliveries rejected should be measured and assessed as to their impact, causes, and cost.

A: 13-3   Important methods of improving quality include:

- Top management commitment, encouragement, and support

- Finding out and ensuring that the needs and wants of the customer are met

- Simpler organization

- Standardization, simplification, and specialization of outputs

- Doing the right thing the first time

- Prevention of errors and defects

- Improved labor relations, communication, teamwork, and doing the work with a positive attitude

- Training and retraining of all staff

A: 13-4   Important aspects of quality in personnel work include proper communication, cooperative teamwork, quick action on requests, unbiased treatment of all staff, fairness, friendliness, helpfulness, and reducing and eliminating redundancy and red tape where possible.

A: 13-5   Productivity improvement actions should be tackled one at a time, prioritizing those actions that are likely to have the greatest impact and can be implemented as soon as possible.

A: 13-6   Having key sources for all supplies in each classification may be advantageous; nevertheless, one should continually review competitive and new suppliers because it has been found useful for a buyer to see all salespersons to avoid being too dependent on past suppliers.

A: 13-7   In physical distribution, some of the customer's requirements include:

- Accuracy of all transactions as to item, number, size, price, and so on

- Dependability of delivery

- On-time delivery

- Rapid and honest response to order inquiries

- Quick correction of errors

A: 13-8　Accounts receivable can be reduced by:

- Prior checking of the credit rating of customers

- Following up all overdue accounts immediately

- Setting up systematic formal procedures for following up and collecting delinquent accounts

- Putting one person in charge of the control of all accounts receivable

A: 13-9　In order to improve administration, first you must define the various kinds of administration, for example, financial, personnel, purchasing, production, warehousing, sales, billing, and collection. Above all, it needs to be ensured that the administration activities are necessary. Much administrative work may not be needed for the achievement of added value and should be eliminated. It is important, therefore, to know the proportion of productive work. This examination should be done thoroughly and reviewed from time to time. Administration productivity improvements can then be made by using better office procedures, methods, and systems, including the elimination of unnecessary redundancies.

A: 13-10　In order to identify improved methods of physical distribution, it is useful to keep a record of:

- The SKUs that comprise the bulk of the sales

- The location and distribution of the customers, and the items moved

- Whether the items could be rearranged into more effective and efficient distribution

- The direction and timing of the items' distribution

- The frequency of need for the items at particular locations

- The security required

# 14

# How to Make the Best Use of Your Time, Efforts, Knowledge, and Other Resources

## UTILIZE YOUR TIME EFFECTIVELY

Perhaps the most important requirement of improving productivity is the continual best use of your time, whether you are an entrepreneur, manager, supervisor, or employee. This can not be emphasized enough because time is our scarcest resource, and its proper use is one of the most critical determinants of productivity. Using time effectively helps us complete the important jobs before threatening deadlines are upon us. Time management does not mean working harder, but rather working smarter. Effective time management allocates time systematically and helps to produce much more with less effort. It is very important that managers and supervisors not only practice good time management, but also motivate and train their staff to develop a habit of making the most effective use of their time.

Following are examples of methods that have been found to help manage time more effectively and usually bring about big improvements in performance:

- It is exceptionally useful to keep a list in writing, for a few days, of every job one is doing for at least 20 to 25 minutes. Daily activities should be prioritized. The most important tasks should be done first and the unimportant ones should be eliminated.

- The high quality of work must be ensured. Duplications and unnecessary or redundant work should be identified and eliminated.

- All documents should be kept in proper folders and files in order to be easier to retrieve and to minimize the risk of losing them.

- The telephone should be used efficiently.

- The time of managers should be reserved for planning, negotiating, guiding, and controlling, personnel matters, and for tasks that their supervisor asked them to do personally.

- Any work that subordinates can do or for which they can be trained should be delegated to them.

- Idle time should be utilized to improve productivity. It is important to be "productive," not just "busy."

# APPLY PRODUCTIVE METHODS OF ORGANIZATION, WORK DESIGN, AND JOB ASSIGNMENT

Proper organization is very important for productivity. Organizational improvements are often inexpensive, yet may yield surprisingly abundant rewards. Once it has been determined what has to be done, the managers and supervisors should decide how the task at hand is to be divided into elements and how the elements of the task are assigned to individuals or groups of individuals. Each person must know clearly what needs to be done.

The various functional areas, such as planning, design, engineering, purchasing, production or service provision, warehousing, marketing, distribution, and administration, need to be coordinated and their relative functions defined. Excessive administrative vertical layering should be avoided, but horizontal interface between interrelated organizational elements needs to be ensured. The desirable number of subordinates under a supervisor varies from organization to organization depending on the type of activity and the type of employees involved. It seems feasible to supervise about eight people at senior or complex levels. In more routine operations of larger organizations a supervisor may supervise more than 30 employees. The objective is to achieve a simpler and flatter hierarchical organization. Reorganization should be done carefully because change interrupts work and worries employees. Any change should be prepared with advance communication and, where possible, consultation with those involved.

The most effective structuring of an organization also depends on its type, the organization's products or services, resources, processes used, and the kinds of customers or clients. Over time, these may change. For example, when a product matures, some of the production engineering department can be reassigned to introduce redesigned components, which can bring about higher productivity and lower labor or capital (machinery, equipment, and so on) costs.

Balancing the use of full-time and part-time workers can significantly increase productivity and reduce cost. Critically important operations need more full-time workers to ensure stable, high-quality work, but overtime costs can be minimized by using part-time workers at regular rates at peak periods. If certain services are not required regularly enough to utilize staff fully, such services may be outsourced when needed.

An example can be found in the lodging industry. Hotels and motels consist of several departments, such as accommodations, food service, beverage service, and miscellaneous operations. These need to be coordinated so that each assists the performance of the others. A considerable part of labor in hotels and motels is in occupations that tend to be sensitive to changes in seasonal and other specific demands. These jobs may include clerical and advertising occupations. In other occupations, such as room cleaning personnel, food and beverage service workers, and security staff, minimum staff needs to be maintained all the time. Seasonal fluctuations in this industry can be offset by building up the business of conventions and meetings because conventions and meetings often take place in winter when the tourist business is reduced.

Productivity in the hotel and motel industry can benefit from such organizational changes as operating economies achieved by the use of convenience foods, and participating in chains that can take advantage of direct bookings and finding sponsoring organizations. Franchising can be more efficient than independent operations because individual franchisees can be guided and assisted by centralized administrative offices. The latter also can afford advanced managerial and operational techniques, provide nationwide advertising, access to favorable credit, and the benefits of volume purchasing of supplies.

The various types of organizational elements and layers should be shown on organization charts that indicate the functions to be performed by each unit, and sometimes by individuals, as well as their clear reporting relationship. It is important that everyone in the organization report to one person in order to avoid divided responsibility. Individual positions need detailed job descriptions.

The specific authorities and responsibilities of each position need to be defined clearly. Overlapping, unnecessary elements and duplications should be eliminated. It is desirable to locate decision making at the lowest possible level of hierarchy, nearest to the job affected by the decision. Value-added work, that is, work spent on actual production or service for the customer, should be clearly distinguished from non-value-added work, such as getting ready, obtaining components, waiting, and so on. The time gained by minimizing non-value-added work can be utilized for improving product and service quality and saving expenses.

Several related jobs may be combined into more comprehensive jobs, which can make the employee see the effects of his or her work on the service to the customer or client. For example, in insurance, an employee can be made responsible for issuing policies, adjustments, billing, claim settlements, and so on. This can ensure that the customers are provided with consistent, efficient, and high-quality service by one worker. In another example, the job of bank tellers can be merged with customer service functions that can be performed when there is no client waiting line. This helps to avoid staff sitting and working at their desks while customers are waiting in line for tellers. Customers and clients like a "one-stop shopping" arrangement that enables them to deal with the various aspects of their problem at one central point. Employees who are made responsible for comprehensive, flexible jobs need to be trained in all aspects of the combined function.

The right person should be selected for the job. The productive utilization of staff requires the selection of staff so as to enable the best use of their potential, but employees should not be assigned duties clearly beyond their capabilities. Employees should be expected to complete their job satisfactorily, but should be allowed to plan the details of their work. Successful completion of a task leads to a feeling of accomplishment. Labor input costs can also often be minimized by utilizing employees in various functions according to time availability. While job content may be made more flexible and comprehensive, however, quality work also requires continuity so that workers, and in particular supervisors, engineers, and other professionals, are not transferred to new assignments too frequently, nor in the midst of another job.

In order to ensure productivity and quality improvement, organizational effectiveness needs to be reviewed from time to time from both the administrative and technical viewpoints. Considerable improvement in the productivity and quality of output, as well as cost savings, can be achieved by improving methods and processes with a view to ensuring that the best use is made of both human and physical resources. Process improvements can reduce the number of defects, errors, waste, and outdated and redundant production and administrative work that is no longer needed. Methods and techniques should be regularly reviewed and updated in order to ensure that the most productive methods are used.

It is important to ensure, however, that the changes introduced for productivity and quality improvement systems are not excessively time-consuming. When new methods are developed, the setting up of tests or methods laboratories may be helpful. Companies have successfully implemented such laboratories by setting up simulated trial runs of new methods of operation.

# CONTROL IS NECESSARY FOR ACHIEVING PRODUCTIVITY

Organizations perform successfully and productively only if progress is kept under continual review and, if necessary, performance is adjusted from time to time to ensure that organizational objectives and targets are met using the most efficient methods. Otherwise, time and other resources would be wasted. Generally, "too good" and "too bad" should be eliminated. The first may be too expensive while the second may not satisfy the customer.

In the control process, managers and supervisors are responsible to ensure that the planned, scheduled, and initiated actions are progressing as intended. They must be aware when problems or bottlenecks arise, interpret the information, determine whether corrective action is needed, seek clarification when necessary, and take corrective actions as required.

The controls should be reasonable and simple. They also should be economical and not cost more than the consequences of no control. The control should be balanced, that is it should not place undue emphasis on one part of the desired result because that could distort performance. Control must be timely to enable the correction of errors or faulty products before the error or fault is repeated over and over again and major damage is done. For this reason, it is also important to make a decision on a corrective action quickly and implement the action expeditiously.

If possible, the control should be comprehensive, that is, placed at points where several operations can be controlled at once, instead of controlling each operation individually. For this purpose the controls should be built in at important strategic points of the operation where the work is done and where the impact is the greatest. For example, major high-tech companies reported substantial productivity improvement by building control points into the operation along the assembly line so that immediate corrective action could be taken when necessary. Further, the more precisely the planners have stated the goals, objectives, and standards of the organization, the less checking is necessary.

Still, various types of controls are required depending on the type of organization. The main control areas include labor, manufacturing overhead, administration, materials, operations, inventory, management (including financial statements, performance, and other productivity ratios), and environmental controls. Although most of these control areas need no further explanation here, a few examples are provided.

Production overhead costs that need to be controlled include all costs other than direct labor, materials, and energy. Production overhead costs consist of administration related to manufacturing production or service output, indirect labor, clerical and sales supplies, insurance, taxes, indirect materials, heating, air conditioning, water, light, and other utilities, waste disposal and pollution control, as well as depreciation of physical workplace and fixed assets.

The data that are useful for productivity and profitability measurement and control purposes include such ratios as return on assets (operating profit/operating assets), profit margin (operating profit/sales), and turnover of assets (sales/operating assets), production costs over sales volume of production, operating expenses over sales, as well as real productivity ratios (those expressed in volume ratios) such as number of chairs produced per employee, per employee-hour, per floor area, and/or per volume of material or energy used.

An example from the lodging industry illustrates how controls can be implemented by calculating productivity ratios and comparing them to targets. The sales by waiters can be divided by the hours worked by waiters. The resulting sales per hour should meet the targeted ratios. In food operations, portion control is one of the important elements of cost control. More-detailed discussion of control through measurement was presented in Chapters 3 through 7 of this book.

The control of input costs can also be illustrated by another example from the lodging industry where, say, in the food department it starts with food purchasing. It is necessary to decide what items are needed, in what quantities, and where satisfactory items can be bought at the lowest prices. Insufficient quantities lead to customer dissatisfaction, while too large quantities tend to lead to extra capital cost, spoilage, and shrinkage. Establishing standards and formal specifications helps to maintain stable, high-quality food at minimal cost.

Inventories of stocks should be taken at least monthly and reconciled with what should be there after the recorded sales. Larger organizations may track inventories as often as daily, and smaller units may take inventories weekly, for instance, if the owner/manager is away frequently. Discrepancies should be investigated and their causes eliminated. The areas where stocks are stored should be locked, and access to them limited to authorized personnel.

The quality and quantity of foodstuffs received need to be inspected and tested at least on a sample basis. Specific individual(s) need to be placed in charge of receiving and be held accountable for having performed the prescribed duties. Invoices should be reconciled with orders, discrepancies investigated, and their causes corrected. Care needs to be taken to avoid waste in the preparation of food and to control portion sizes in order to ensure that they are in accordance with the predetermined costs and prices of the various menu items. The actual costs should be regularly verified against the targets.

The techniques of control follow from the objectives and standards established when the plans and schedules are set up. Useful types of standards include volume or quantity standards, quality or condition standards, cost standards, and time per person and per process standards. Results are measured and compared against objectives and standards, and corrective actions taken when warranted. Overall controls are based on the organization's profit and loss statement, inventory levels, and productivity ratios. These can be compared with past performance and expressed in terms of variances from the plan.

For the purpose of monitoring progress, accurate production or service records are required, including work orders, bills of material, and so on. Obviously, not every single product or service provided can be checked. Two useful work-saving methods include sampling and control by exception. Sampling was discussed separately in Chapter 4. Control by exception demands a report from the checkpoint only when the process deviates from the plan. For major expenditures or lengthy jobs, it may be necessary to conduct formal reviews from time to time.

As most operations involve work by a number of employees, it is useful to devote part of each staff meeting to discussing the progress of the work, general performance and productivity improvement, problem areas, outstanding results, as well as unsatisfactory performance. Examples of various measures used for control purposes include:

- Assignments completed on time

- Rejection rates

- Wrong product or service delivered

- Error in count, price, terms, or physical aspects of the transaction

- Quality not meeting design or manufacturing/service specifications

- Jobs per labor hour, day, or week

- Actual output per machine compared with the standard output per machine

- Percent of jobs run incomplete due to shortage of material

- Percent of schedule changes made because of tooling or machine unavailability

- Percent of total inventory that is obsolete

- Percent loss of materials due to shrinkage and pilferage

# PERSONNEL CONTROLS AND ACCOUNTABILITY

Employee performance should be adequately monitored and evaluated on a regular basis. Employees will know what is expected of them if they are provided with simple, clear, and specific job descriptions that emphasize what and how much is needed. In specifying what is expected, the quality requirements also must be specified. The staff must be provided with the necessary training for each of the functions they might be required to perform. The employees should also be informed of specific requirements, if any, and given milestones by which they must meet their targets. If they don't, it should be determined why not, and the employees should be provided with guidance for improvement. The purpose of control is not to criticize the performers but to ensure that the required processes are performed. Employees should not be held responsible for meeting production standards if the prescribed methods have built-in inefficiencies. For example, employees can not perform work if the needed components are not available to them.

The controls should not stifle creativity. Performance measures and controls will affect performance because the employee will endeavor to meet the requirements of the measures and may neglect other aspects of performance. Productive organizations of today prefer self-control, meaning that each employee must make sure that his or her own performance meets or surpasses the required standard. Each should be shown how they themselves can evaluate their performance, how to report problems, or resolve the problem by themselves.

It has been observed that increases in the frequency of inspections are rated rather low among the methods of quality improvement. Constant attention by the workers to do the job right in the first place is the most important quality control and quality enhancer. It is necessary to identify and communicate to the workers information on, for example, design deficiencies, likely possibilities of defects or errors, and common defects. The development of effective self-checking methods tends to result in a reduction of defects, and to bring substantial results. It has been found useful to set aside time each week to discuss observations regarding rejects and errors, as well as suggestions of how improvements could be implemented in the near future.

Productivity plans and measurement systems are of little value unless they are used, and the productivity efforts of the organization are not taken seriously unless they involve accountability. People should, therefore, be held accountable for their assigned responsibilities, including their expected productivity. It is essential to build responsibility for performance into all jobs, regardless of organizational level, difficulty, or skill required. Managers will endeavor to improve productivity only if their responsibility for productivity improvement is clearly defined and they are held accountable for productivity performance. Accountability can be achieved by defining the missions and tasks clearly and specifying the expected performance and productivity results for various measured activities in their work assignments or job descriptions. Clear, reliable, and timely productivity measures need to be developed and the expectations compared with actual performance.

Employees need measures to know their performance. It is important to ensure that the data are valid and meaningful, fair, and not too cumbersome. The data to be collected should focus on significant outputs. Useful guidance on possible and desirable employee performance targets can be acquired from various available statistics. In hotels and motels, for example, the number of rooms with a bath that can be serviced by a full-time equivalent room service worker per day depends on layout and walking distances, which vary from resort hotels and bungalows to high-rises. In the food department, data indicate the average amount of food and/or beverage sales per seat per month or per year, or per waiter/waitress.

Feedback on performance needs to be provided to employees. Each employee must know whether the quantity and quality of his or her work is satisfactory and, if not, why not and how to do better next time. Performance evaluation and appraisal are needed to ensure that productivity improvement action is taken and the requirements are implemented. Proper performance evaluation is done to ensure that the valid demands of customers or clients are met, that the employee's good performance is recognized and rewarded, and that the poorer performers are shown how to improve. The main purpose of performance evaluation is to show employees how to improve rather than to criticize. It is necessary, however, to inform the worker if his or her performance does not meet expectations.

Formal performance appraisal meetings between the employee and his or her supervisor are usually held once a year, but it is highly desirable to evaluate work informally throughout the year in order to enable immediate and continual improvement. Performance should, therefore, be discussed in the course of normal work, and improvement guidance should be provided all along. In the appraisal form, productivity targets need to be included in quantifiable form.

Performance interviews should be preceded by the preparation of a written fact sheet about the employee. Points to be covered may, among others, include the quantity, quality, and timeliness of work performance, reliability, attendance, cooperation with coworkers, customer or client service satisfaction, changing work habits, and other points that may be relevant to the employee's general and productivity performance.

Attendance may become unsatisfactory. Absenteeism may appear as an increased frequency of missed days or half-days of work or increased unjustified absences on Mondays or Fridays. Poor attendance may also appear in frequent lateness to work, extended lunch periods, or often leaving work early. Personnel problems may appear in the form of irritability, sudden untidiness, or noticeable changes in normal behaviour. The employee may become unreliable. He or she may handle assignments in an unusually sloppy way or be inclined to put things off unnecessarily. Performance problems may mean hazards to health and safety. These may be caused by careless handling and maintenance of equipment. Such factors may lead to frequent accidents, injuries, and lost time. If there is a need for correction of performance, the supervisor should indicate what corrective steps must be taken.

One of the main requirements of providing quality products and services to customers and other users is the prevention of defects and errors. We have already noted that prevention is usually much less expensive than correction. It leaves time for more productive matters and certainly results in better customer satisfaction and loyalty. The prevention of defects and errors needs to be implemented throughout the entire production process, from raw material and components purchasing through product design, production, delivery, and administration. Quality control data need to be gathered and evaluated continually throughout the process, with inspections carried out as necessary.

Process control charts should be received together with incoming material, and the purchase should be fully inspected if indicated by specific variance. When defects are observed by the workers, they are often authorized in modern plants to stop the assembly line to find the cause of the defect and eliminate it right away. In administration and related customer services successful organizations have reported that they routinely review and audit all invoices before sending them out in order to avoid customer dissatisfaction.

In order to prevent errors and defects, the workers need to know exactly what to do and how to do it. If they do not, they need to be shown, trained, or given access to the answers. It is very important that they should not be afraid to ask, and to report a problem that they can not resolve themselves. Innovation, product redesign, and the development of new products or services, as well as continually seeking new applications for current products and services, have become more and more important characteristics of successful organizations in recent years. These developments provide natural opportunities for eliminating typical causes of defects and errors. They also point to poor characteristics of design or process that can be altered to produce a more defect-free process or product.

# HOW TO BENEFIT FROM TECHNOLOGICAL PROGRESS

The use of current technology and appropriate capital investment are important factors of productivity both in the private and public sector. New technology can make operations faster, more accurate, easier, safer, as well as less expensive. It may enable operations that earlier were impossible, and make it possible to transfer some of the work to the customer, for example, in banks through ATMs (automated teller machines) or online banking. With the help of computers, well-trained employees can deliver many services to the consumer and do so better, faster, and cheaper.

New technologies have opened up many ways of improving the productivity and quality of products and services, often at much lower cost. For example, new materials have not only made automobiles lighter but also more durable and more resistant to rust. Product designs that have been well planned and tested and use common components have eliminated many setup and retooling tasks and costs.

Wireless and screw-less designs have eliminated many possible causes of inefficiency and quality problems. CAD/CAM (computer-aided design/computer-aided manufacturing) technologies have reduced retooling problems, enabled the radical shortening of changeover and setup times, and helped make immediate process adjustments as soon as required. The new technologies have made smaller batch sizes feasible, with the result that production can be better tailored to customer requirements. The possibility of effective small-batch production has enabled smaller enterprises to compete more effectively in all markets. Looking at progress from the other side, quality improvement often requires the use of new technologies and new equipment, although most companies do not consider this kind of expenditure a primary factor of quality improvement.

Advancements in information technology (IT) have led to doing more with less and doing better with less, for example, through improved and more rapid planning, control, and communication. There is no end to examples of possible applications of computers. They can be used to identify operational bottlenecks, analyze service performance or layout difficulties, improve and speed up communication with suppliers and customers, as well as enable the maintenance of lower levels of inventories of raw materials, goods-in-process, and finished goods. In repair services, for instance automobile service shops, computers assist in locating, analyzing, and correcting defects much faster and more accurately than ever before.

The technological improvements don't have to be high-tech or consist of major breakthroughs or be implemented in the form of very expensive equipment. They may consist of unextraordinary developments that are part of the day-to-day production system. For example, in hotels and motels, no-iron laundry, nonslip bathtub mats, throwaway dishes, vending machines, shoe cleaning machines, modern cooking and sanitation equipment, and direct dialing from guest rooms reduce the time needed by staff and help enhance productivity. At the same time they increase guest convenience and satisfaction. Another example of simple improvement is to place adequate signs, which are very important to those who are looking for a particular facility. Seeing an attractive and promising sign can also attract new customers. The signs must be clear, legible, visible, neat, and attractive.

It is, therefore, necessary to keep abreast of possible simple improvements, as well as technological developments, encourage innovation, and use the latest technological tools if they are warranted. One should always keep an eye open for new machinery, equipment, and materials that could replace the current ones, improve productivity and profitability, and support continuing improvements in the operations. Further productivity improvements can also come from the new opportunities and knowledge the new equipment enables. It is useful to have someone made responsible to keep abreast of new technologies.

It is very important, however, to analyze and justify—preferably with statistical evidence—the need for new equipment before it is acquired. The analysis needs to be thorough, examining all aspects of the operation. First it must be established that the operation itself is needed and that it is efficiently organized. It must be determined what the system is supposed to accomplish, whether the existing equipment is not still reasonably up to date, and whether it is not capable of meeting the requirements of the actual work.

It also should be examined whether the equipment's capacity is properly utilized, underutilized, or overutilized. It has often been found that better use of existing equipment and technology can, in itself, lead to considerable productivity improvement without major expenditures. If the current workload does not ensure full capacity utilization, you may want to sell the old equipment or acquire new equipment on a rental basis for the time it is required. The selling price of the old equipment should also be taken into consideration. It needs to be determined whether the proposed equipment is suited to do the job and whether its future utilization is more efficient and effective than that of the present system.

The analysis may indicate that new equipment is not needed and that, for instance, the proposed automation would, in fact, be more expensive than to continue performing the work, perhaps manually, with the old system. Savings achieved by avoiding unnecessary equipment purchase costs may reduce the enterprise's financial requirements. The latter may be an important consideration, particularly for smaller enterprises, whose biggest problem tends to be a lack of finances.

In acquiring new equipment, the human factors, such as safety and working conditions, including light, air, and noise, are among the most important aspects to consider. It is also important to ensure that qualified staff will be available to operate the new equipment, and that the staff will be adequately trained to properly use the new equipment. The productive effects of information technology require that its users operate it properly without being handicapped by an unnecessary flow of information or hardly understandable technical instructions. The future users of the new equipment should, therefore, be involved in its justification and selection.

There are also institutional and organizational elements to be weighed. For example, the relationship between the length of a production run and the time needed for changeover to another product or service may vary from industry to industry and from country to country because the relative cost and scarcity of various factors of production may bring about different answers. For example, if labor is relatively cheap and capital is relatively expensive, then using more labor would be preferable. The capital expenditure needs to be justified by cost–benefit analysis, including a determination that the new equipment to be purchased is the best alternative to do the job. Otherwise unnecessary and misdirected investment may take place. Excessive or unnecessary purchase of fixed assets is a typical problem of many enterprises, particularly smaller ones.

Improvement in productivity through proper equipment management requires the efficient and full utilization, regular maintenance, and repair of all equipment and facilities. The useful life of equipment and facilities can be greatly enhanced by a proper preventive maintenance and calibration program. The ready availability of necessary spare parts ensures that the machines in question will not stay idle until the necessary spare parts are obtained.

Unnecessary equipment should be sold, or mothballed if future use can be foreseen. Studies have shown that the best utilization of plants, equipment, and other resources tends to be more important than the relative value of the assets. To achieve this objective, one should keep an up-to-date inventory of assets, including their current valuation and utilization. All interested employees of an organization should be told of the availability and capability of the organization's physical facilities.

It must, however, always be kept in mind that new technologies do not always lead to the expected positive results, and may even have negative side effects. Studies have shown, for instance, that the worth of some of the advances in information technology (IT) may be questionable, particularly if compared to the huge capital sums invested. The side effects may include stress on the operating personnel, or higher reject rates and downtime because of the lack of sufficient experience. These side effects are probably higher in the initial period of using the new equipment because faults may have to be corrected, adjustments required, time is needed by the operators to learn the new operation, or because of a lack of knowledge of how to exploit the full capacity of the new asset. It should also be kept in mind that technical advances may often increase the workload rather than increase productivity, such as when people are enabled to reach you virtually all the time with their e-mail, cell phone, or telephone messages, or hamper you with data overflow. Technical possibilities have not always led to practical use because the technical advances were too rapid to allow enough time for practical implementation, or because new technologies have been introduced without thorough prior analysis and the required staff training. Negative side effects can cause indirect costs of new equipment. These costs must not be underestimated by managers.

Many of the above findings may seem obvious, but experience shows that they are too often neglected. If properly addressed, the above improvements can significantly enhance productivity.

# QUESTIONS

Q: 14-1  What are some of the requirements of an effective job assignment?

Q: 14-2  How can you reduce waste?

Q: 14-3  What production overhead costs need to be reduced?

Q: 14-4  How can you improve materials productivity?

Q: 14-5  List some of the causes of labor turnover and absenteeism.

Q: 14-6  Before deciding to buy new equipment, what do you have to determine about the old equipment?

# SUGGESTED ANSWERS

A: 14-1  Some of the requirements of an effective job assignment are:

- A clear job description should be provided to each employee.

- Productivity and quality objectives need to be included in all job descriptions.

- Each employee should have a clear understanding of how his job fits into the overall organization and what his or her work contributes to the organization.

- Each employee should know to whom he or she reports.

A: 14-2  You can reduce waste by identifying duplication and backtracking, for example, through observing excessive expenses when comparing them to estimates or standards. You can also reduce waste by examining work-flow charts.

A: 14-3  Overhead costs that can be reduced include all costs other than direct labor, material, and energy; therefore, overhead includes costs such as administration, indirect labor, sales, office supplies, insurance, utilities, waste disposal, pollution control, as well as depreciation of physical workplace and fixed assets.

A: 14-4  Materials productivity can be improved, for instance, by keeping careful records of:

- Materials purchased

- Direct and indirect materials actually consumed during the period, and comparing usage to a standard or target

- Price paid for materials per unit of output

A: 14-5  Some of the causes of turnover and absenteeism are dissatisfaction with compensation, objectivity and treatment, the lack of rewards and incentives, poor communication, little or no training or career development, poor guidance, and inadequate help from supervisors.

A: 14-6  Before buying new equipment, you need to establish:

- Whether the old equipment can still do the job
- Whether and how the new equipment would better meet your needs
- Whether it would be cheaper or otherwise more advantageous to buy new equipment
- What indirect costs the purchase would involve in terms of quality, reject rates, inventory, staff overtime, and so on
- Whether there is a less expensive way to achieve the desired results

# Endnotes

## Preface

1. Adam Smith, *The Wealth of Nations* (Edinburgh, 1776).
2. Caroll D. Wright (the first United States Commissioner of Labor), *Hand and Machine Labor* (Washington, DC: 1898).
3. Imre Bernolak, "Canadian Sources of Industrial Productivity Measures and Some Comments on Measurement Methods," in *Productivity Measurement* III (Paris: Productivity Measurement Advisory Service, Organisation for Economic Cooperation and Development, 1966).
4. Imre Bernolak, "Is Growth Obsolete," comment in *The Measurement of Economic and Social Performance: Studies in Income and Wealth* 38 (New York: National Bureau of Economic Research, 1973).
5. Imre Bernolak, "The Whole and Its Parts: Micro and Macro Productivity Research," in *Dimensions of Productivity Research* II (Houston: American Productivity Center, 1980).
6. Imre Bernolak, "Interfirm Comparisons in Canada," in *Productivity Measurement: An International Review of Concepts, Techniques, Programmes and Current Issues,* David Bailey and Tony Hubert (eds.) (Westmead, Farnborough, Hants, England: Gower for the British Council of Productivity Associations, 1980).
7. Imre Bernolak, "The Measurement of Outputs and Capital Inputs," in *Productivity Measurement: An International Review of Concepts, Techniques, Programmes and Current Issues,* David Bailey and Tony Hubert (eds.) (Westmead, Farnborough, Hants, England: Gower for the British Council of Productivity Associations, 1980).
8. Imre Bernolak, "Using Performance Comparisons: Some Policy Implications," in *Productivity Measurement—An International Review of Concepts, Techniques, Programmes and Current Issues,* David Bailey and Tony Hubert (eds.) (Westmead, Farnborough, Hants, England: Gower for the British Council of Productivity Associations, 1980).

9. Imre Bernolak, "New Productivity Thrust from Effective Measurement," in *1981 Spring Annual Conference and World Productivity Congress* (Detroit, MI: American Institute of Industrial Engineers, 1981).

10. Imre Bernolak, "Conventional Wisdom Is Not Always Wise," in *Productivity Measurement and Analysis: New Issues and Solutions* (edited by Imre Bernolak) (Tokyo: Asian Productivity Organization, 1983).

11. Imre Bernolak, editor, *Productivity Analysis and Projections in Selected Key Areas in Asian Countries* (Tokyo: Asian Productivity Organization, 1987).

12. Imre Bernolak, book review of *Productivity Measurement: A Practical Handbook* by Joseph Prokopenko, International Labour Office, Geneva, published in *Optimum,* Bureau of Management Consulting, Supply and Services, Canada 1989/90, vol. 20-1.

13. Imre Bernolak, "Linking Managerial Action to Productivity Measures," in *International Productivity Journal* III (Washignton, DC: International Productivity Service, 1991).

14. Imre Bernolak, "Productivity from the Pusztas to the Prairies," in *Europe Productivity Ideas* (EPI) 2 (Brussels: European Association of National Productivity Centres, 2002).

## Chapter 7

1. Imre Bernolak, editor, *Productivity Measurement and Analysis: New Issues and Solutions* (Tokyo: Asian Productivity Organization, 1982).

2. Imre Bernolak, editor, *Productivity Analysis and Projections in Selected Areas in Asian Countries* (Tokyo: Asian Productivity Organization, 1987).

3. The Centre for InterFirm Comparison, CIFC Ltd., 32 St. Thomas Street, Winchester, Hampshire SO23 9HJ, United Kingdom.

## Chapter 10

1. Edward M. Coates, III, "Profit Sharing Today: Plans and Provisions," *Monthly Labor Review* (Bureau of Labor Statistics, 1991).

2. The Profit Sharing/401k Council of America (PSCA) *49th Annual Survey of Profit Sharing and 401(k) Plans* (Chicago: Profit Sharing/401k Council of America, 2006).

3. David L. Wray, President, Profit Sharing/401k Council of America (Chicago, IL), phone interviews February 22 and March 30, 2007.

4. The Profit Sharing/401k Council of America (PSCA) *49th Annual Survey of Profit Sharing and 401(k) Plans* (Chicago: Profit Sharing/401k Council of America, 2006): page 9 and Table 12.

5. Carl F. Frost, John H. Wakeley, and Robert A. Ruh, *The Scanlon Plan for Organizational Development: Identity, Participation, and Equity* (East Lansing, MI: Michigan State University Press, 1974).

# Chapter 13

1. ISO, 1, ch. de la Voie-Creuse, Case Postale 56, CH-1211 Geneva 20, Switzerland www.iso.org

# Index

## A

acceptable quality level (AQL), 63–64
acceptance sampling, 63–64
accountability, productivity measures
    and, 43
accountability controls, 222–25
administration, opportunities in,
    200–204
American Productivity and Quality
    Center (APQC), xiii
American Productivity Center (APC),
    xiii
analysis,
    in identifying productivity
        problems and opportunities,
        37–115
    of methods of operation, 47–49
Asian Productivity Organization
    (APO), 196
attendance, and productivity, 129, 224
attribute quality characteristics, 14
average, 53
awareness training, 132

## B

benchmarking, 23–24, 48
    benefits of, 107–15
    performance improvement
        opportunities revealed by,
        114–15
British Institute of Management, 108
budget justification, productivity
    measures in, 42
business plan

common elements of successful,
    181–82
how to develop, 181–85

## C

Canadian Interfirm Comparison
    Program, 9
capacity planning, 183
capital, misuse of as barrier to
    productivity, 124
capital input measures, 78–79
Centre for InterFirm Comparison, 108
change, resistance to, 122
clerical staff, improving productivity
    of, 203–4
clients, improving communication
    with, 135–36
common cause variation, 57
communication, opportunities for
    improving, 135–40
comparative organizational
    performance analysis,
    importance of, 107–10
computers
    in productivity training, 170
    use in productivity improvement,
        226
consultants
    role in SMEs, 24
    role in training, 171–72
control
    measures used for, examples,
        221–22
    necessity of for achieving
        productivity, 219–22

control chart, 57–58
controls
    accountability, 222–25
    personnel, 222–25
corrective action, gathering facts
    before taking, 39–49
cost of quality, 12
costs, and quality, 10–13
critical path method (CPM), 186,
    188–89
cross-sectional productivity analysis,
    26
customer complaints, 72
customer feedback, 71–72, 127–28
customer orientation, 127–28
customers, improving communication
    with, 135–36

**D**

data
    needed for measuring output in
        services, 75–76
    presentation of, 53–55
    steps in gathering essential, 62–64
dispersion, 54
distribution, 54
    physical, productivity
        improvement in, 209–11
distribution plan, 185

**E**

effectiveness-type measures, 6–7
efficiency-type measures, 6–7
effort, making best use of, 215–28
employee performance
    measures, 223
    monitoring and evaluating, 222–24
employee stock ownership plans
    (ESOPs), 150
employees
    communication with, 135–36
    effective management of, 128–30
energy productivity, 59–60
equipment management, for
    productivity improvement,
    227–28

equipment productivity, 59

**F**

feedback, customer, 71–72
financial incentives, 148–50
financial performance ratios, 111–14,
    220
firefighting services, productivity
    improvement in, 34
401(k) plans, 149
frequency distributions, 54–55
functional literacy, of employees, 162

**G**

gainsharing, productivity, 151–55
Gallatin, Albert, 148–49
Gantt chart, 186, 187–88
goal setting, productivity measures
    in, 41

**H**

human factors, importance to
    productivity, 128–30, 227

**I**

in control process, 57
incentives, financial,
    as motivators, 148–50
incentives, nonfinancial,
    involvement and participation as,
        146–48
    as motivators, 144–46
indirect workers, 29–30
    productivity improvement ideas
        for, 32–35
information technology (IT), in
    productivity improvement,
    225–26, 228
input, measuring, 67–79
input costs, control of, 220
input indexes, calculation of, 90
inputs

affecting quality, 198
    measurement of, 77–79
integrated organizational performance
    analysis, 110–14
interfirm productivity comparisons,
    xvii–xviii, 60, 107–8
International Organization for
    Standardization (ISO), 195
inventory management, productivity
    improvement in, 207–9
inventory ratios, 56
involvement, as motivator, 146–48
ISO 9000 quality standards, 195–96

## J

job assignment, applying productive
    methods of, 216–19
job interviews, manager's duties, 161
jobs, and quality, 10–13

## K

knowledge, making best use of,
    215–28
knowledge workers, 29–30
    measurement needs of, 31–32
    productivity improvement ideas
        for, 32–35

## L

labor input measures, 77–78
labor productivity, 7
lot tolerance percent defective
    (LTPD), 64

## M

management
    importance to productivity,
        128–29
    productivity training for, 167
    role of, in productivity
        improvement, 130–32
managers

changing role of, 130–32
    of SMEs, and measurement of
        results, 24–26
marketing subplan, 184
Marshall Plan, 107
mean, 53
measurement
    in identifying productivity
        problems and opportunities,
        37–115
    of input(s), 77–79
    of output, 67–79
    of results, importance, 24–26
measures
    for control purposes, examples,
        221–22
    effectiveness-type, 6–7
    efficiency-type, 6–7
    of productivity, 4–5
median, 53
meetings, conducting productive, 137
methods of operation, analysis of,
    47–49
mode, 53
motivation, importance to enhancing
    productivity, 143–44
multifactor productivity measures,
    8, 60

## N

negotiations, successful, preparing
    and conducting, 138–40
negotiators, characteristics of
    effective, 139–40
network management techniques,
    188–89
nonfinancial incentives
    as motivators, 144–46
    involvement and participation as,
        146–48
normal distribution, 54

## O

objective quality characteristics, 14
objectives, need for setting, 177–80
operating assets, 100

operating plans, need for formulating,
    177–80
operating profit, 110
operations
    high-quality, how to design,
        193–211
    productivity measures in, 43
oral presentations, preparing effective,
    136–37
organization, applying productive
    methods of, 216–19
organization, improving
    communication within, 135–36
organizational performance analysis
    comparative, importance of,
        107–10
    integrated, method of, 110–14
out of control process, 57
outcome, versus output, 6
output
    measuring, 67–79
    versus outcome, 6
    in services, data needed for
        measuring, 75–76
output indexes, weighted, calculating
    over time, 90
output indicators, criteria and
    methods of choosing, 67–70
output quality, identifying problems
    and opportunities of, 71–74

# P

Pareto analysis, 56–57
partial productivity measures and
    concepts, 7–8
participation, as motivator, 146–48
people
    importance in negotiations, 139
    importance of to productivity, 125
people-oriented skills, 132
performance
    interviews, 129
    problems, 129–30
    versus productivity, 3
performance analysis, comparative
    organizational, importance of,
        107–110

performance improvement
    opportunities, revealed by
        benchmarking, 114–15
    productivity measures in, 42–43
personnel controls, 222–25
personnel subplan, 184
physical distribution, productivity
    improvement in, 209–11
planning
    importance of to success, 177–89
    productivity measures in, 41
plans
    key characteristics of successful,
        180–81
    types of, 182
plant organization, productivity
    improvement in, 207
population, 53
presentation of data, 53–55
process consultants, 172
process control, 55–58, 63, 64
process control chart, 57–58
product design, 204–5
production
    ensuring high-quality, 197–200
    versus productivity, 3
production management, improving
    productivity of, 201–3
productivity
    barriers to, 121–24
    calculating, of two organizations
        with heterogeneous products
        using ULR weights, 86–90
    comparison of in two
        organizations with
        heterogeneous products,
        93–94
    corrective actions to improve,
        165–66
    definition, 3–15
    disincentives to, 123–24
    history of, xvii
    importance of to SMEs, 21–24
    improving through organization
        and technology, 175–228
    interfirm comparisons, xvii–xviii
    most important lessons in, 164
    versus performance, 3
    versus production, 3

and quality, 4
and standards, 5–8
and technology, 3–4
understanding, 1–36
productivity analysis, 25–26
  as a training tool, 43
  general approaches to, 43–47
productivity calculations, tasks for
  practicing, 95–105
productivity changes, calculation
  over time when outputs are
  homogeneous, 83–84
productivity consultants, role in
  training, 171–72
productivity differences, calculating,
  81–83
productivity gainsharing, 151–55
productivity improvement, 8–10
  basic elements of, 119–72
  benefits of, 9–10
  criteria, 193–94
  ideas for professionals and
    knowledge workers, 32–35
  setting priorities for, 193–94
  special needs of smaller
    enterprises, 21–35
  steps for SMEs, 26–28
productivity improvement programs
  past weaknesses, 124–27
  successful, lessons from, 121–32
productivity indexes
  calculating, from weighted output
    and input indexes, 91–93
  comparing in two hospitals,
    81–105
  construction from output and
    input indexes when outputs
    are homogeneous, 85
  construction from raw productivity
    data, 83–84
  methods of construction, 81–94
productivity indicators, 39–47
productivity measurement
  general approaches to, 43–47
  lack of, as barrier to productivity,
    124
  principles and requirements,
    58–62
  training, 166

productivity measures, 4–5
  developing, 53–64
  effectiveness-type, 6–7
  efficiency-type, 6–7
  labor productivity, 7
  multifactor, 8
  partial, 7–8
  total factor, 8
  training on, 166–67
  use of indicators, 39–47
productivity opportunities,
  identifying, 37–115
productivity problems, identifying,
  37–115
productivity ratios, 220
productivity training, as special need,
  163–68
productivity trend analysis, 26
professionals
  productivity improvement ideas
    for, 32–35
  special productivity needs of,
    28–35
profit sharing, as motivator, 148–50
profits, and quality, 10–13
program evaluation and review
  technique (PERT), 186, 188
progress, technological, how to
  benefit from, 225–28
proof of quality, 194–97
purchasing, productivity improvement
  in, 206–7

**Q**

qualitative indicators, 61–62
quality
  definition, 13–14
  high, effects of, 10–13
  inputs affecting, 198
  meaning of, 13–15
  measurement of, 55–58
  output, identifying problems and
    opportunities of, 71–74
  poor, effects of, 10–13
  and productivity, 4
quality characteristics, 14
quality circles, 148

quality improvement programs, successful, lessons from, 121–32
quality indicators, 39–47
quality of work life, as motivator, 145
quantitative indicators, 61–62

# R

range, 54
Rationalisierungs Kuratorium der Deutschen Wirtschaft (RKW), 107
reports, preparing effective, 136–37
research and development (R&D), in output measurement, 75–76
resource consultants, 171–72
resources, making best use of, 215–28
results
measurement of, importance, 24–26
motivating for, effective methods, 143–55
retail stores, productivity improvement in, 34–35
retraining
as motivator, 145–46
need for continual, 159–72
reward programs, 144–45

# S

sales, and quality, 10–13
sales value of production, 110
sampling, 63
Scanlon, Joseph, 183
Scanlon plan, for productivity gainsharing, 153–55
scatter, 54
scatter diagram, 55
scheduling, in productivity improvement, 185–89
security, workplace, 145
service, ensuring high-quality, 197–200
services
data needed for measuring output in, 75–76
special productivity needs of, 28–35
simplification, in product and service design, 204–5
skill diversification, 165
skill training, 132
skill-based workers, 30–31
small and medium-size enterprises (SMEs)
importance of productivity to, 21–24
special productivity needs of, 21–28
steps to improve productivity of, 26–28
Smith, Adam, xvii
special cause variation, 57
specialization, in product and service design, 204–5
standard deviation, 54
standardization, in product and service design, 204–5
standards
in product and service design, 204–5
and productivity, 5–8
statistical concepts, basic, 53–55
stock keeping units (SKUs), in inventory management, 208–9
strategic plans, need for formulating, 177–80
subjective quality characteristics, 14
supervisors, changing role of, 130–32
suppliers
improving communication with, 135–36
involvement in quality assurance, 165

# T

task-oriented skills, 132
teamwork, as motivator, 146–48
technological progress, how to benefit from, 225–28

technology
  as barrier to productivity, 124
  and productivity, 3–4
time
  making best use of, 215–28
  utilizing effectively, 215–26
total factor productivity (TFP)
  measures, 8, 60
total quality management (TQM),
  197
training
  evaluation of, 167–68
  as motivator, 145–46
  necessity of for success, 159–63
  need for continual, 159–72
  in productivity, as special need,
  163–68
training methods, effective, 168–70
training needs, criteria for defining,
  160
transportation costs, incoming,
  reducing, 206

# U

unit labor requirement (ULR)
  weights, 88–89
  calculating productivity of
  two organizations with

heterogeneous products
  using, 86–90
United States Department of Labor,
  xvii, 107

# V

value added, 59
variables quality characteristics, 14
variance, 54

# W

Watts, Alan, xiii
work design, applying productive
  methods of, 216–19
work in process (WIP), 62
workplace security, 145

# X

$\bar{X}$ chart, 57–58

# Z

zero defect strategies, 12

# Belong to the Quality Community!

Established in 1946, ASQ is a global community of quality experts in all fields and industries. ASQ is dedicated to the promotion and advancement of quality tools, principles, and practices in the workplace and in the community.

The Society also serves as an advocate for quality. Its members have informed and advised the U.S. Congress, government agencies, state legislatures, and other groups and individuals worldwide on quality-related topics.

## Vision

By making quality a global priority, an organizational imperative, and a personal ethic, ASQ becomes the community of choice for everyone who seeks quality technology, concepts, or tools to improve themselves and their world.

## ASQ is...

- More than 90,000 individuals and 700 companies in more than 100 countries
- The world's largest organization dedicated to promoting quality
- A community of professionals striving to bring quality to their work and their lives
- The administrator of the Malcolm Baldrige National Quality Award
- A supporter of quality in all sectors including manufacturing, service, healthcare, government, and education
- YOU

**Visit www.asq.org for more information.**

# ASQ Membership

Research shows that people who join associations experience increased job satisfaction, earn more, and are generally happier*. ASQ membership can help you achieve this while providing the tools you need to be successful in your industry and to distinguish yourself from your competition. So why wouldn't you want to be a part of ASQ?

## Networking

Have the opportunity to meet, communicate, and collaborate with your peers within the quality community through conferences and local ASQ section meetings, ASQ forums or divisions, ASQ Communities of Quality discussion boards, and more.

## Professional Development

Access a wide variety of professional development tools such as books, training, and certifications at a discounted price. Also, ASQ certifications and the ASQ Career Center help enhance your quality knowledge and take your career to the next level.

## Solutions

Find answers to all your quality problems, big and small, with ASQ's Knowledge Center, mentoring program, various e-newsletters, *Quality Progress* magazine, and industry-specific products.

## Access to Information

Learn classic and current quality principles and theories in ASQ's Quality Information Center (QIC), *ASQ Weekly* e-newsletter, and product offerings.

## Advocacy Programs

ASQ helps create a better community, government, and world through initiatives that include social responsibility, Washington advocacy, and Community Good Works.

**Visit www.asq.org/membership for more information on ASQ membership.**

*2008, The William E. Smith Institute for Association Research

# ASQ Certification

ASQ certification is formal recognition by ASQ that an individual has demonstrated a proficiency within, and comprehension of, a specified body of knowledge at a point in time. Nearly 150,000 certifications have been issued. ASQ has members in more than 100 countries, in all industries, and in all cultures. ASQ certification is internationally accepted and recognized.

## Benefits to the Individual

- New skills gained and proficiency upgraded
- Investment in your career
- Mark of technical excellence
- Assurance that you are current with emerging technologies
- Discriminator in the marketplace
- Certified professionals earn more than their uncertified counterparts
- Certification is endorsed by more than 125 companies

## Benefits to the Organization

- Investment in the company's future
- Certified individuals can perfect and share new techniques in the workplace
- Certified staff are knowledgeable and able to assure product and service quality

Quality is a global concept. It spans borders, cultures, and languages. No matter what country your customers live in or what language they speak, they demand quality products and services. You and your organization also benefit from quality tools and practices. Acquire the knowledge to position yourself and your organization ahead of your competition.

**Certifications Include**
- Biomedical Auditor – CBA
- Calibration Technician – CCT
- HACCP Auditor – CHA
- Pharmaceutical GMP Professional – CPGP
- Quality Inspector – CQI
- Quality Auditor – CQA
- Quality Engineer – CQE
- Quality Improvement Associate – CQIA
- Quality Technician – CQT
- Quality Process Analyst – CQPA
- Reliability Engineer – CRE
- Six Sigma Black Belt – CSSBB
- Six Sigma Green Belt – CSSGB
- Software Quality Engineer – CSQE
- Manager of Quality/Organizational Excellence – CMQ/OE

**Visit www.asq.org/certification to apply today!**

# ASQ Training

## Classroom-based Training

ASQ offers training in a traditional classroom setting on a variety of topics. Our instructors are quality experts and lead courses that range from one day to four weeks, in several different cities. Classroom-based training is designed to improve quality and your organization's bottom line. Benefit from quality experts; from comprehensive, cutting-edge information; and from peers eager to share their experiences.

## Web-based Training

### Virtual Courses

ASQ's virtual courses provide the same expert instructors, course materials, interaction with other students, and ability to earn CEUs and RUs as our classroom-based training, without the hassle and expenses of travel. Learn in the comfort of your own home or workplace. All you need is a computer with Internet access and a telephone.

## Self-paced Online Programs

These online programs allow you to work at your own pace while obtaining the quality knowledge you need. Access them whenever it is convenient for you, accommodating your schedule.

### Some Training Topics Include
- Auditing
- Basic Quality
- Engineering
- Education
- Healthcare
- Government
- Food Safety
- ISO
- Leadership
- Lean
- Quality Management
- Reliability
- Six Sigma
- Social Responsibility

**Visit www.asq.org/training for more information.**